Open Access

GIS in e-Government

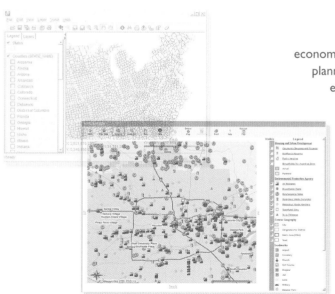

economic development
planning and zoning
environmental monitoring
law enforcement
property assessment

R.W. Greene

ESRI PRESS
REDLANDS, CALIFORNIA

ESRI

Open Access: GIS in e-Government

ISBN 1-879102-78-1

First printing April 2001

Printed in the United States of America.

Library of Congress Cataloging-in-Publication Data
Greene, R. W.
    Open access : GIS in e-Government / R. W. Greene.
        p.   cm.
    ISBN 1-879102-87-0
    1. Internet in public administration—United States.
    2. Geographic information systems—United States.
    I. Title.
JK468.A8 G74 2001
352.7'45—dc21   2001002069

Published by ESRI, 380 New York Street, Redlands, California 92373-8100.

Books from ESRI Press are available to resellers worldwide through Independent Publishers Group (IPG). For information on volume discounts, or to place an order, call IPG at 1-800-888-4741 in the United States, or at 312-337-0747 outside the United States.

Contents

Preface   vii
Acknowledgments   ix

Introduction   1

1   **Mapping an e-government future   7**
Thanks to GIS, the San Diego region's e-government capabilities are both varied and deep, giving us a glimpse of what shape a fully formed e-government future may take.

2   **Ready for work   23**
Interactive mapping helps lower the unemployment rate in Delaware by showing job seekers exactly where their needs can be met.

3   **Northwest e-passage   31**
A noteworthy economic renaissance in one Puget Sound city—no, it's not Seattle—has been aided by interactive mapping.

4   **OzarkIMS   39**
In Arkansas, the Razorbacks' hometown waited for the latest technology to put together its first interactive GIS site, and the results take its e-government effort way downfield.

5   **Property, parcels, and plots   47**
In one Ohio county, there is virtually no information about a piece of property you can't find online.

6 **E-votes, e scores** **55**
Arizona is one of the first places to try Internet voting; it's also where one county's interactive site is helping citizens unravel the tangled skeins of political jurisdictions.

7 **Extending the Pulaski** **63**
The disastrous fire season of Summer 2000 results in a nation-sized wildfire GIS application that lets fire managers hundreds of miles from firelines make accurate assessments of what resources are needed.

8 **Online in Oregon** **71**
Interactive mapping solutions using GIS are not limited to big-budget urban agencies. A rural Oregon county used a small demonstration grant to put up a Web site that shows off the area's economic possibilities.

9 **A healthy place to call home** **79**
The U.S. Department of Housing and Urban Development combines the vast data resources of the federal government to create an interactive site that can help citizens in any community.

10 **E-city, e-community** **87**
You name it, you can find it on Sacramento's interactive mapping Web site: trash pick-up days, streetcar routes, the most graffiti-plagued neighborhoods, property values, and much more.

11 **Worth a look** **95**
Interactive mapping can show Web visitors much more than routine government matters, as these snapshots show.

12 **Geography Network and e-government** **105**
What lies ahead for e-government?

Other books from ESRI Press

Preface

Geographic information systems have been an integral part of the infrastructure of technologically savvy government agencies and institutions for decades. From wide marble corridors in Washington, D.C., to local city halls, GIS has been making a difference. One reason is that GIS efficiencies cover such a broad spectrum of activity, mirroring the broad constituency that governments must serve. Forward-thinking governments have found GIS applications essential for the internal management of utilities, law enforcement, emergency response, health care, transportation, the environment, and a host of other responsibilities.

Now the Internet revolution is transforming institutions, and in fact, bringing new ones into existence: there's not just mail anymore, there's also e-mail; there's not just commerce anymore, there's also e-commerce; and of course, governments are becoming e-governments, serving constituents electronically in ways unimaginable only a few years ago.

Combining the efficiencies of GIS with the efficiencies of e-government is a natural integration. It means bringing those internal GIS efficiencies out to the external world.

When GIS is served over the Internet, e-government becomes much more than an online way to help constituents fill out forms, or register to vote, or pay their property taxes. As commendable as those new online capabilities are, constituents using the interactive mapping power of GIS over the Internet can do much, much more.

Internet mapping lets them combine geographic layers, such as zoning and tax-incentive areas, to see—for themselves—what parts of a city are best suited for establishing a new business. With easy-to-use drop-down lists and check-off boxes, they can verify—for themselves—whether rumors of a rising crime rate have any validity. They can see—for themselves—how the demographics of the city are changing, where the jobs are moving, where the most environmentally hazardous parts of town are located.

This past year, the Internet portal known as Geography Network℠ has begun incorporating many of these innovations on a national and international scale, helping agencies and constituents alike to leverage their current GIS investments many times over.

GIS and e-government, by helping people see things for themselves, are bringing new dimension and new energy to the Information Age.

Jack Dangermond
President, ESRI

Acknowledgments

This book could not have been written without the generosity of the IT professionals and GIS specialists in the government agencies profiled in these pages. Their names are listed at the end of each chapter, but all deserve an extra round of applause here for supplying so much accurate information and so many patient explanations, not to mention maps and suggestions—and above all, enthusiasm for their work. There's a reason why the agencies profiled here are at the head of the pack in the e-government marathon; these folks are that reason.

Much appreciation also goes to many at ESRI for their own feats of patient pedagogy, particularly David Cardella and Bernie Szukalski. Many thanks also to Steve Hegle, Gary Amdahl, Tim Craig, Kasey Quayle, Rick Ayers, Todd Rogers, Brian Harris, Kim Burns–Braidlow, Clem Henriksen, Jeff Baranyi, Tim Clark, Charlie Magruder, Paige Spee, Pat Cummens, Wojtek Gawecki, Sarah Osborne, Russ Johnson, Christopher Thomas, Eileen Napoleon, Judy Boyd, and Nick Frunzi.

Central to making the book a reality were Christian Harder, ESRI Press manager, who really wrote the template for it with his own *Serving Maps on the Internet;* Michael Hyatt, who designed the pages and did the production and copyediting; Jennifer Galloway, who designed and produced the cover and handled image editing; and Michael Karman, who edited the manuscript.

Introduction

No matter where you're standing or sitting as you read this, you're within the territory of some government agency, and probably several of them: some city, or county, or state, or economic incentive zone, or vector control district. No matter what your political persuasion—or your cynicism about political persuasions—geography is one of the fundamental systems by which we have chosen to govern ourselves.

There's also a revolution in government services going on around you as you read this. Commonly known as e-government, the delivery of government services to the public through the Internet and other digital means is transforming the relationship between governments and the governed.

Applying for permits, searching for documents, paying taxes—all those things you used to have to do by standing in long lines, you can now do sitting at your desk, over the Internet.

A frictionless interaction is now within reach, diminishing, if not ending, a traditionally adversarial relationship.

Virtually every town, county, state, and provincial agency on the continent is working on some kind of e-government service. At the federal level alone, the Web portal known as FirstGov (www.firstgov.org) offers a one-stop shop to five hundred million Web pages and twenty-seven thousand Web sites offering information and assistance.

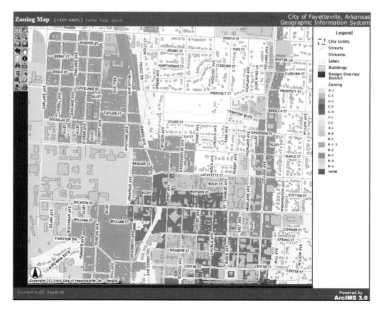

Making things clear with GIS

Taking the lead in e-government service delivery are agencies like the ones profiled in this book.

They are finding that connecting geography—in the form of interactive maps—to the e-government process makes it work better. Geographic information systems (GIS) technology, which many are already using, lets them offer services with qualities rarely associated anymore with analog government—speed, efficiency, flexibility.

The reasons why are simple, but no less powerful because of that fact.

E-government services based on maps and GIS can show residents a simple diagram of a city's council districts, or give them online tools and data for detailed forecasts of regional growth. Whether basic or complex, these tasks are based on our ability to see spatial relationships in the world.

Maps are an easier and more natural way of looking at that world. Maps are not forms to be filled out. They are not pages and pages of boilerplate language

that only faintly resembles English. They are not disembodied voices telling you to push 1 for yes and 2 for no.

Maps are inherently intuitive, easy to understand and use—especially in an age when the visual has been exalted as the premier form of communication.

It's been possible for years, of course, to put up static maps of the city on a municipal Web site. But Internet Map Server (IMS) technology adds all the power of GIS to Internet service delivery.

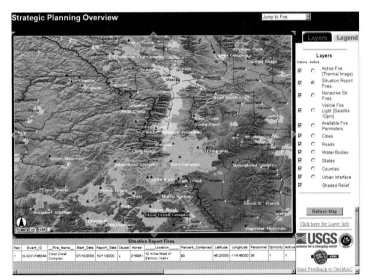
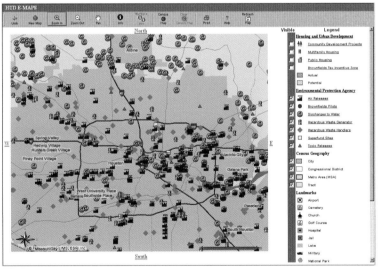

A page from the GeoMAC Web site, left, lets firefighters keep tabs on wildfires and plan strategy over the Internet. A Department of Housing and Urban Development (HUD) page, right, lets visitors see the relationship between population and environmental hazards.

Online, not in line

That means that constituents can create digital supermaps on their home PCs, interactive maps that link data of almost any kind or size to a particular geographic location.

How does this happen?

Visualize GIS as a number of two-dimensional transparencies, or layers, each containing objects in geographic relationship with each other. A layer of streets is an obvious example of such a layer; a street map is the kind of layer people are most used to seeing and understanding.

With a GIS, you create different layers of geographic data that you want—all the police stations in your city, for example, or all the burglaries, or all the city council districts. A GIS then lets you place these layers on top of each other. Since you can see all the information in each layer at the same time that you see the information in all the other layers, you can see relationships that might never have occurred to you before—especially if you're used to looking at tables of numbers.

You might place a layer of police station locations over a layer of locations of burglary reports, and also a layer of city council districts. Then, perhaps for the first time, you could see that five police stations in one city council district means few burglaries, while another city council district that has only two stations has 75 percent of all burglaries.

As more people have come to understand the value of this kind of analysis, the number of GIS applications, in both the private and public sector, has exploded in recent years.

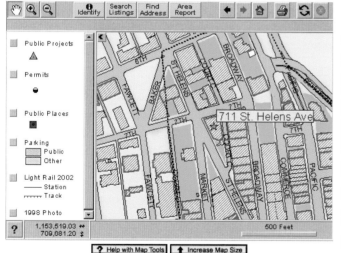

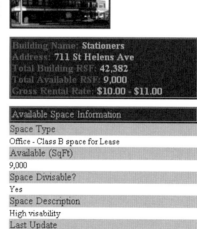

The city of Tacoma, Washington, uses interactive mapping as a key element in its economic development process. Visitors to the Tacoma Web site can get answers to questions about what buildings are available for sale or lease all over the city, and also get important details about price, parking, and availability.

Part of a revolution

One reason is that there is virtually no limit to the kind of information you can place in a GIS layer, as long as it has a geographic reference point: demographic trends, soil types, income levels, voting tendencies, poverty rates, pollution levels, epidemics, cereal brand preferences, high school drop-out rates, college scholarship rates, and television-watching preferences—the list is limited only by the imagination.

In addition to the power of this unlimited data, GIS incorporates tools to analyze these relationships in many different ways, and it can create models of these analyses.

With IMS technology, ESRI helps government agencies bring these tools into living rooms and home offices worldwide, letting visitors to an agency Web site measure distances, create buffer zones, pinpoint addresses, or find the best routes to school and to work.

In the commercial world, GIS has created powerful tools for marketing, distribution, sales and sales forecasting, customer service, and other disciplines where geography is intrinsic to the process.

Many expect that the economic payoffs of e-government efficiencies will be on the same order of magnitude as those that have accrued to private enterprise during the digital revolution—payoffs that for some have been astronomical.

Efficiency is not the only benefit. It is not even at the top of the public's list of the improvements they expect e-government will bring to their lives. A survey by the Council for Excellence in Government found that the e-government benefit people think will be most useful is its ability to let them hold government more accountable—to keep an eye on what government is doing.

Many observers say it is hard to overstate the long-term effects the e-government revolution may have on our lives, on the interaction between governments and the governed.

"The online revolution is as much about reinvigorating our democracy as it is about improving services and operations," says Patricia McGinnis, president and CEO of the Council for Excellence in Government.

Put another way, when citizens can see for themselves—with IMS—that burglaries are much worse in one part of town than in another, it increases the chances that something might get done about it.

Interactive mapping also increases flexibility in a community. Changing conditions of any kind—in the local weather or economy, for example—can be mapped and put up on a Web site immediately for the benefit of citizens.

Left, an innovative site in Delaware lets job seekers find employers, child care, and transportation. Right, aerial photos aid economic development efforts in Oregon.

Issues of concern

Agencies are still feeling their way along the hallways of this new kind of service delivery. Some issues, such as security, are as much a concern to public agencies as they are to private ones. Some others, such as privacy, are a particular worry.

Property tax maps, for example, are becoming a common application on many IMS e-government sites. Because a GIS can show information down to street and parcel level, many agencies are still working out the best ways to balance the public's right to know public information with the privacy rights of the average homeowner. Some of these property tax sites, for example, will let you see the name, address, and tax bill of every homeowner in town, while others restrict access to information about officials such as police and judges; still others hide all names.

Liability issues are also still being worked out. Most e-government Web sites require visitors to read and accept privacy and liability agreements before they are allowed to click through to the public information on the rest of the site.

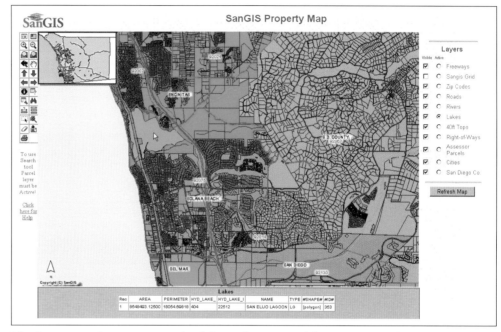

The San Diego County region of California has some of the most comprehensive Internet mapping services available on the continent. The latest interactive software technology from ESRI, ArcIMS, powers the interactive property map, above.

Sowing the seeds

ESRI has long recognized the benefits that will accrue to society from a wider dissemination of GIS, especially over the Internet. The architecture of its newest and most robust technology to date, ArcIMS®, is designed to let agencies build entire interactive mapping Web sites easily; to help them integrate their earlier IMS investments; and to make use of geographic data they have accumulated over years and even decades.

The agencies featured in *Open Access: GIS in e-Government* range from small rural organizations using basic desktop GIS software, to federal agencies with powerful mapping services that let you figure out how many children live within five miles of a Superfund toxic-waste site.

As a representative cross-section of IMS use in e-government, this book is designed to sow the seeds for even greater benefits to even more citizens.

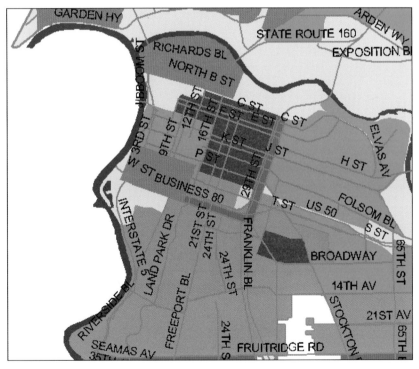

The city of Sacramento's mapping site gives visitors a clear view of such mundane but necessary municipal matters as trash pick-up schedules. Putting this kind of information on the Web can save an agency many hours and dollars over time, by freeing up employees who would otherwise have to answer such questions over the phone or in person.

Mapping an e-government future

Its vast blue oceanfront alone makes San Diego an area of great wealth. Add blue skies as far as the eye can see and temperatures that rarely fall below 60, and the riches of the San Diego area become almost embarrassingly plentiful. One of its most valuable assets is also one of its least visible: a vast repository of digital information that describes virtually any aspect of the San Diego region that you could want to know— its physical geography, its land uses, its economy, its transportation routes, its people and their possible futures. Like beachfront and skyline, this information is easily accessible to anyone who needs it, through interactive GIS mapping over the World Wide Web. Indeed, the wealth of San Diego's mapping data, its utility, and its ease of use may foretell what the mature e-government system of the future will look like.

A wide spectrum

What sets the San Diego region apart is the sheer comprehensiveness of the mapping applications available over the Web. Many local agencies across the continent are putting property maps on the Web, while others are seeing the efficacy of election and voting applications, and still others like what can be done with crime maps.

You will find all of these applications, and more, in use on a Web site somewhere in the San Diego region. This is because several San Diego public agencies are long-time GIS users, and so have years of data and of experience to work with. In addition, their regional vision and cooperative approach to data sharing have helped them leverage individual assets into a far richer regional whole.

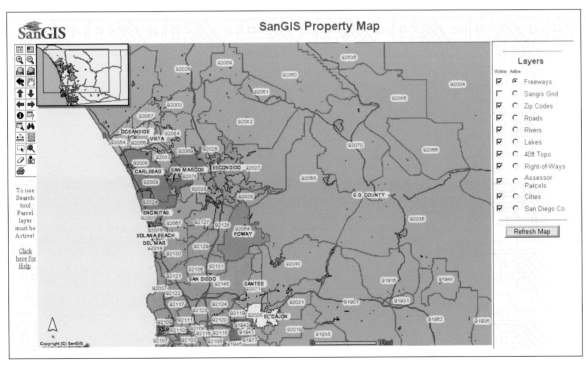

ArcIMS, the newest interactive mapping technology from ESRI, powers this GIS map. Despite the title, much more than property information is available here, as shown by the list of layers to the right of the main map display. To the left of the display are the tools to work with the map: zooming, panning, identifying features, locating addresses, and measuring distance. The small map in the corner shows you which section of the county the larger map shows; this feature can be turned on and off.

A regional perspective

Two organizations that have been critically important to San Diego's success are called SanGIS and SANDAG.

SanGIS, the San Diego Geographic Information Source, is a partnership between San Diego city and county governments. Among other responsibilities, it houses and maintains basemap data layers, and more than two hundred other data layers from other agencies, all covering more than forty-two hundred square miles of county territory. Its data includes information for public safety, planning and development, facilities management, route mapping, and decision support.

SANDAG (San Diego Association of Governments) is a consortium of the county, eighteen municipal governments within the county, and some specialized local, state, and federal agencies. One of the association's important missions is providing information for regional decision making on transportation, economic development, environment, growth, and housing.

The two organizations share data and resources, and sometimes combine forces; they share the same ultimate goal of disseminating geographic information to a wide audience.

The same ArcIMS map, after zooming in on an area north of San Diego. Individual parcels and other features are now visible. After checking the appropriate box from the layer list at the right, a parcel or other feature can then be identified by clicking the button on the toolbar at the left, outlined in red. Here, the Lakes theme has been activated, and information about the body of water at the cursor location is shown beneath the map display.

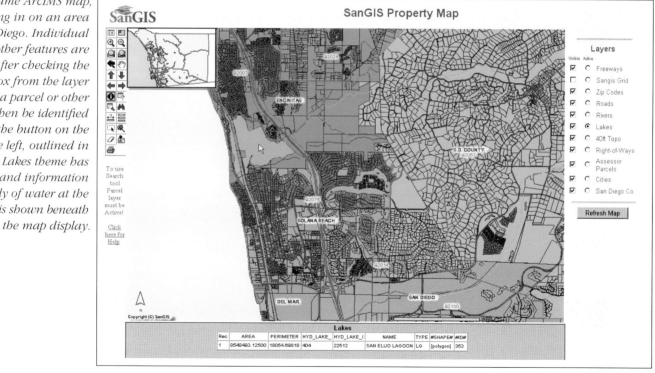

Calming the waters

Just how effective can GIS be for e-government?

An El Niño-caused shift of weather conditions in the San Diego area some years ago prompted federal officials to remind citizens they needed to know how close their homes were to flood zones, and whether they would need flood insurance.

Panicked by this message, residents flooded the Public Works Department with calls for maps—three hundred a day, compared to the usual handful of daily requests. The department was overwhelmed.

But within days, SanGIS had put those flood maps up on its Web site. That stanched the tide of calls to the overloaded department, yet made the information easier for residents to get.

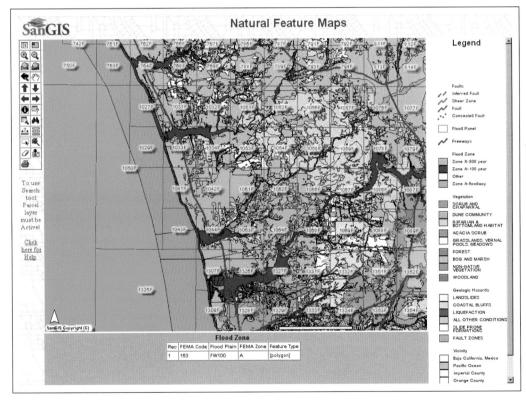

The SANGIS Natural Feature Map, another ArcIMS application, shows the bewildering variety of the natural forces that underly the San Diego region. A property parcel layer becomes visible at smaller scale, allowing users to see if their property is in a flood zone. The grid and grid numbers refer to standard federal flood panel maps.

Shifting economic sands

Both SANDAG and SanGIS have reputations for technological innovation. When the Internet was still in early adolescence, SANDAG decided to make it the primary vehicle for its mission of keeping the public informed of growth patterns, demographic trends, and similar information it views as crucial to good decision making.

SANDAG's economic development focus—providing information to attract employers to the San Diego region, and to support those already established there—was reenergized by the end of the Cold War, an event that meant the San Diego region's defense-based economy would have to change if it were to continue its robust growth.

While the military has hardly disappeared a decade later, the shift to a more diversified economy, with an emphasis on high-tech manufacturing and services, is unmistakable. SANDAG's information and research resources are designed to support exactly such transitions.

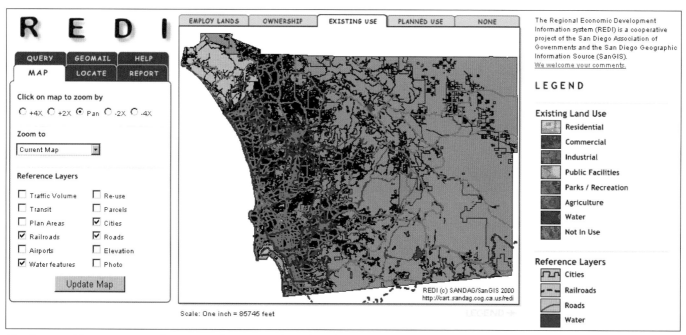

The Regional Economic Development Information (REDI) Web site, a project shared by SANDAG and SanGIS, provides twelve different map layers—the check boxes at the left—that can be overlaid on basemaps selected from the tabs directly above the map. Employ Lands shows lands most suitable for employment activities, while Ownership refers to public lands. The None tab allows the user to work solely with the map layers, without any underlying basemap.

Layers upon layers

The Regional Economic Development Information (REDI) Web site, deployed jointly by SANDAG and SanGIS, contains some of the richest information offerings in the San Diego e-government enterprise. Four different layers of basemap information can be combined with twelve layers of features to answer almost any question about the economic possibilities and land-use potential for any part of the San Diego region.

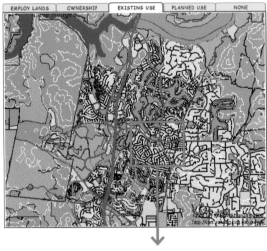

Reference Layers

- **33** Average Daily Traffic (thousands)
- Re-use zones
- Local transit
- Express transit
- Rail transit
- Parcels
- Community Planning Areas
- Cities
- Railroads
- Roads
- Airports
- 4,200 Elevation contours (200')
- Water

Left, existing land use near Interstate 15 in the Rancho Bernardo area of the county. Residential, commercial, and industrial areas are clearly discernible, and contour lines show the steepness of the terrain, a critical factor for developers. When you zoom in more closely on the same area, bottom left, information about average traffic counts on individual roads appears, as do layers showing individual parcels, and an aerial photo. Bottom right, a view of the same general area as the top view, but with land classified differently—by its ability to support high employment levels, as defined in the legend below. Since this area is mainly residential, a use that by definition does not support employment, only the commercial and industrial land around the freeway offramps is highlighted in this view of the area.

Scale: One inch = 756 feet

LEGEND

Employment Lands
- Available Immediately
- Available Long Term
- Planned or Proposed
- Developed
- Unmarketable

Scale: One inch = 3026 feet

REDI for business

Powered by ESRI's MapObjects® IMS technology, the REDI Web site also lets you dig below the surface of its land-use maps down into the wealth of data in the SANDAG Regional Information System.

From its interface, you can focus your search for land that meets certain criteria, or choose an area and create a report.

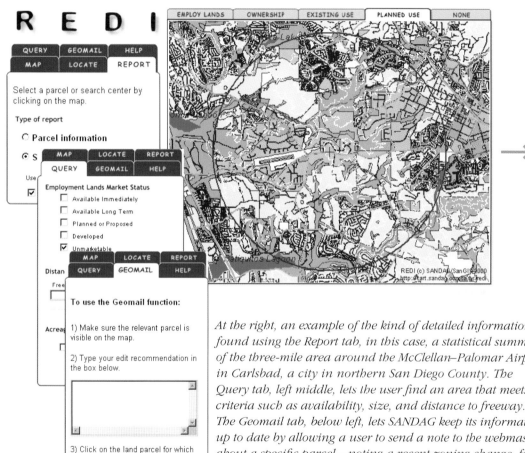

At the right, an example of the kind of detailed information found using the Report tab, in this case, a statistical summary of the three-mile area around the McClellan–Palomar Airport in Carlsbad, a city in northern San Diego County. The Query tab, left middle, lets the user find an area that meets criteria such as availability, size, and distance to freeway. The Geomail tab, below left, lets SANDAG keep its information up to date by allowing a user to send a note to the webmaster about a specific parcel—noting a recent zoning change, for example—directly from the map interface.

Customizing information

SANDAG's Demographic and Economic Mapping System Web pages can answer an infinite number of questions about the people and economy of an area, now and in the future.

1 Using data from SANDAG's Regional Information System, you first select one of the four dat   a sets at the top—Census, Estimates, Forecasts, and Industrial Clusters—then choose which of six administrative layers on which the data set information should be displayed. You then choose which data to map from drop-down lists, 2, and then combine that data with other data, using one of several arithmetical operators, also from a drop-down list, 3. Below, 4, a simple example. Using the Forecast data set for the Major Statistical Area administrative layer, population data from 1995 and projected population data for 2005 have been chosen as variables. The subtraction operation performed results in an equal-interval map that shows which areas of San Diego County are likely to get the greatest population growth in that 10-year period.

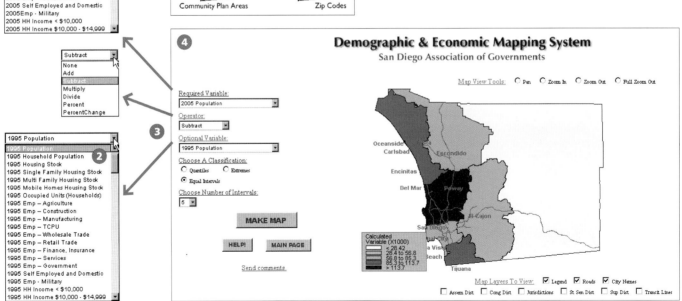

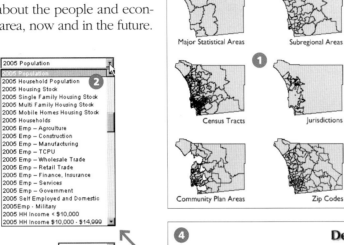

Where the jobs are

The Industrial Clusters mapping tool, developed by SANDAG and the San Diego Regional Technology Alliance, emphasizes the way jobs and industries are distributed around the San Diego region. It serves several groups, among them business owners who need to know where complementary—or competitive—businesses are currently located. Job seekers can use the same mapping tool to see where in the county a job search might be most productive.

Industrial Clusters describes industry groups that more effectively reflect San Diego's newer, high-tech economy, where companies doing work not even conceived of ten years ago—such as Web design—are now important players. The Industrial Clusters system groups companies into categories that describe an end product or service, such as biotech or entertainment, and includes all the businesses involved. This contrasts with the traditional system of assigning businesses to sectors according to function or size, with vague names such as "manufacturing" or "construction."

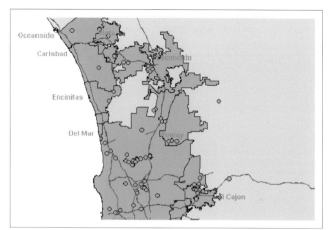

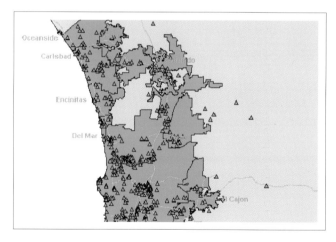

The San Diego region's most important industrial clusters can be found on the Web page by clicking the appropriate box, then refreshing the map. Clicking one of the points on the display will bring up detailed information about the firm. These two maps show how high-tech firms, bottom, are now far more numerous than traditional defense-oriented companies, top.

One map, many votes

While detailed portraits of future population demographics and industrial zones may be invaluable for researchers, for interactive e-government mapping applications to make a difference, they must reach a larger audience, folks who want easy, understandable access to things they need to know on an ordinary day. SanGIS helps with this goal with its ArcIMS application that maps election results across the region.

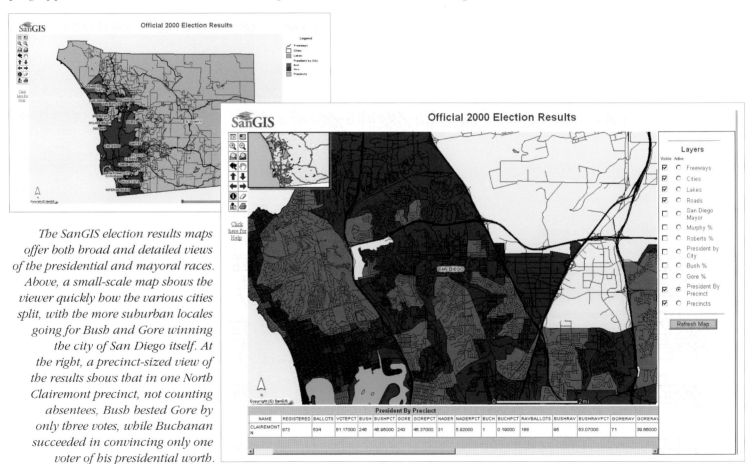

The SanGIS election results maps offer both broad and detailed views of the presidential and mayoral races. Above, a small-scale map shows the viewer quickly how the various cities split, with the more suburban locales going for Bush and Gore winning the city of San Diego itself. At the right, a precinct-sized view of the results shows that in one North Clairemont precinct, not counting absentees, Bush bested Gore by only three votes, while Buchanan succeeded in convincing only one voter of his presidential worth.

Taking a good look around

Another facet of daily, ordinary life in twenty-first-century America is crime. A more informed citizen is usually a safer citizen—and so interactive crime maps are becoming popular on many e-government Web sites. So it is in San Diego, where a partnership between SanGIS and the Automated Regional Justice Information System, ARJIS, produced an interactive Web page that lets residents pinpoint the locations of violent crimes such as assaults; of drug and prostitution arrests; and of traffic accidents, auto burglary, and drunk driving arrests.

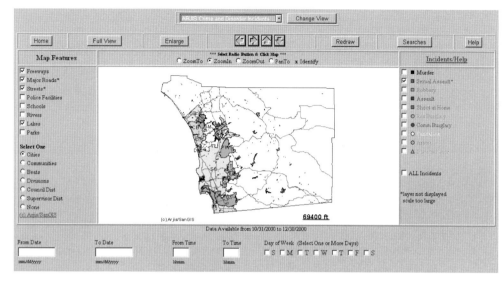

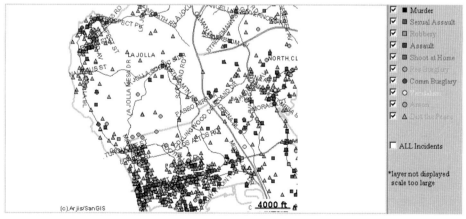

One of the three ARJIS interfaces, above, gives you wide latitude in creating a crime map: which incidents you'll see, what part of the county they're in, and how much other contextual information, such as roads, parks, or schools, to include. At left, an ARJIS crime map shows how one of the region's—and the country's—wealthiest areas, La Jolla, was not immune from crime during a three-month period at the end of 2000.

Mobile mapping

With Californians spending so much time in their cars, interactive Internet mapping of freeway conditions has, not surprisingly, become an important element in San Diego's IMS repertoire. The California Department of Transportation (Caltrans) uses ArcView® GIS software and sensors in the roadway to run its Real Time Speed Map.

The application can be used by commuters getting ready to leave the house or office. Once online, they can see immediately where there are traffic jams—and because it is updated every minute, they can have confidence in its accuracy. Links on the page let them investigate further, to see how bad the delay might be, or whether an alternate route is an option. One

link brings up a list of the actual average speeds on the highway they are interested in. Another links them to a Web site run by the California Highway Patrol, where they can read dispatch center information about a particular traffic accident or breakdown.

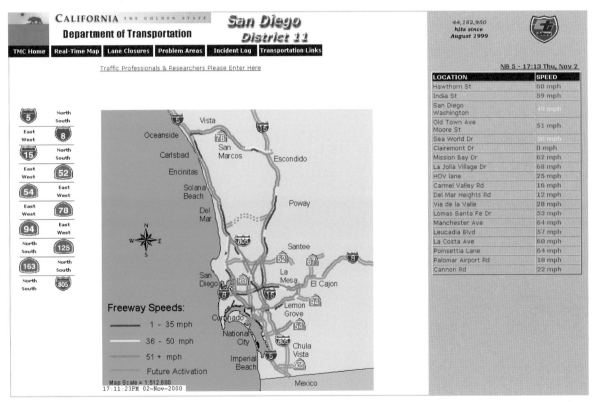

A portrait of the San Diego region freeway system during a typical Thursday afternoon rush hour. Hyperlinks next to each highway symbol bring up an exit-by-exit listing of current speeds. The one to the right of the map display shows conditions on Interstate 5.

No such thing as too much bandwidth
Like being too thin or too rich, it's impossible to have too much high-speed network capacity, especially if you're trying to attract business to your area. San Diego is already one of the most well-wired regions in the country.

With the aid of ArcIMS, the City of San Diego created an interactive map that shows prospective customers just how efficiently business can get done in downtown San Diego.

A close-up view of downtown, right, shows which streets have a high-speed network ready for new businesses to tap into. The network is laid over an aerial photo. Clicking the Identify button brings up the names of digital service providers on a given street. Another link will bring you to the Web page of the company you need.

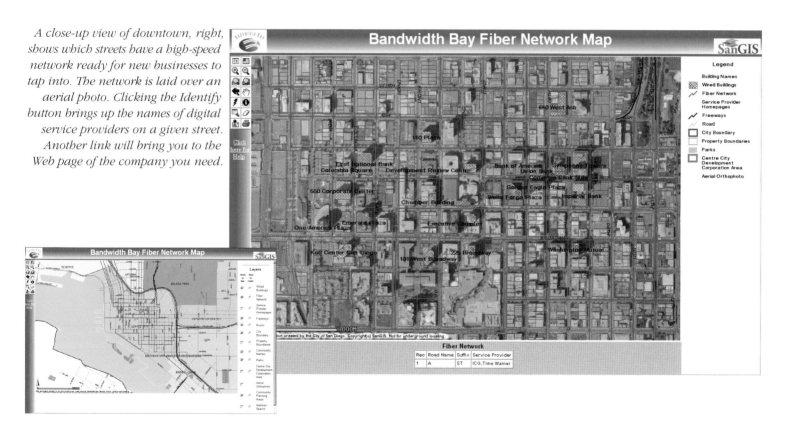

Hands across the border

One of the earliest interactive GIS projects undertaken sought to connect the San Diego region with Mexico.

The San Diego–Tijuana Interactive Atlas, developed by SANDAG and San Diego Dialogue, a nonprofit public policy research center at the University of California, San Diego, provides both U.S. and Mexican census data on a MapObjects IMS interface. It allows researchers to map demographic attributes on both sides of the border— connecting the two countries by letting people view similarities, not merely differences.

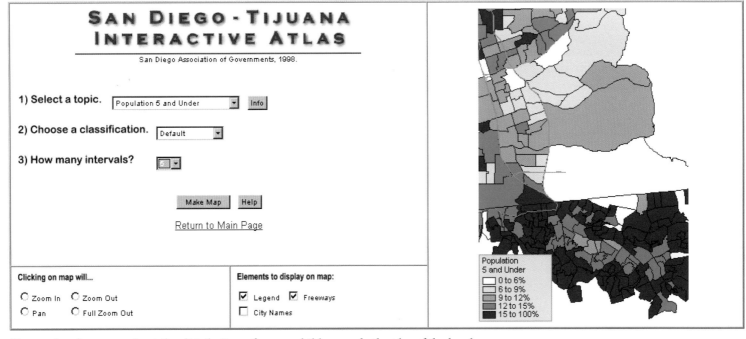

The results of a query about the distribution of young children on both sides of the border are shown at the right. Users can map a variety of attributes, including poverty, education, and housing. The international boundary is the thin black line rising slightly from left to right in the middle of the map.

The systems

SANDAG uses a 450-MHz dual-processor Dell® PowerEdge 1300 with 512 MB RAM and 18-GB RAID-5 disk system. Software includes Microsoft® Windows NT® 4 and IIS 4; MapObjects 2.0 and MapObjects IMS 2.0, and Microsoft Visual Basic® 6.0. An SQL™ Server 7 machine performs statistical aggregations, queries, and reports.

SanGIS uses an 800-MHz dual-processor Compaq® AP750 workstation with 1 GB RDRAM and two 36-GB ultra-wide3 SCSI hard disks. Software includes ArcIMS 3.0, MapObjects 1.2, MapObjects IMS 2.0, Netscape Enterprise Server™ 3.6, Windows NT, and ColdFusion® Server 4.5.

The Caltrans District 11 Real Time Speed Map site uses a PC running Windows NT, as well as UNIX® workstations used as virtual drives. Software includes ArcView GIS 3.1, ARC/INFO® 7.2.1, Avenue™, SnagIT, and NFS Maestro™ from Hummingbird™.

ARJIS hardware includes servers running Sun™ Solaris™ and Windows NT and using MapObjects IMS, MapObjects 2.0, SDE®, Netscape Enterprise Server, Oracle®, Microsoft Visual Basic, and IBM® MQSeries software.

The URLs

www.sandag.org

www.sangis.org

www.arjis.org

www.dot.ca.gov/sdtraffic

Acknowledgments

Thanks to Jeff Tayman and Mark Woodall, SANDAG; Bob Canepa, County of San Diego; Lisa Stapleton and Steve McCarthy, SanGIS; Pam Scanlon, ARJIS; Sandy Johnson, Caltrans.

In a public organization, e-government can smooth administrative processes, reduce the number of committee meetings, and speed decision making. But beyond this streamlining, it is constituents who are the ultimate intended beneficiaries of these new efficiencies. In few places is this so clearly visible as Delaware. The state Department of Labor's Office of Occupational and Labor Market Information built a Web site that uses GIS to help Delaware residents who are looking for work to make their own maps of locations of almost all the state's public and private employers—the people with the jobs. Then they went a step farther, adding tools to the Web site to increase the chances for job-search success, not only for job seekers, but for the state's employers as well.

The 1990s brought big changes to the way American society helped its more disadvantaged members. Discontent with the traditional social service programs known collectively as welfare led to a variety of reforms—ways in which the chronically unemployed could be moved away from a system of direct public assistance, and into education and training programs, and then into jobs. These programs were codified in a sweeping federal reform in 1996 that built on the successes of individual states.

One was Delaware's "A Better Chance Welfare Reform Program" which by one measure had succeeded wildly: the number of people receiving direct assistance fell an astonishing 49.9 percent from early 1994 to 1998.

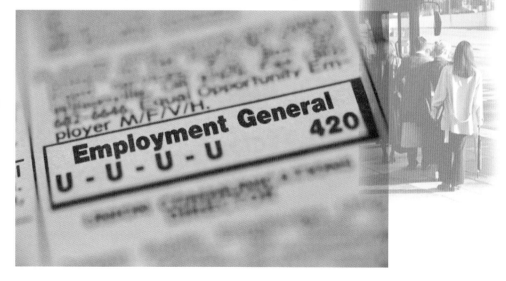

One-stop employment Web shop

The state Department of Labor was one of several state agencies tasked to discuss ways of reforming Delaware's welfare system. Among the goals that Governor Tom Carper assigned the group was to come up with solutions that integrated new technologies, so as to reduce administrative costs.

Disseminating information about jobs and education was at the heart of getting people off welfare, so the Internet became an obvious technological alternative.

At the department's Office of Occupational and Labor Market Information (OOLMI), staff members had already been experimenting with technology, specifically ArcView GIS. The software was ideal for such an application, because GIS can link large amounts of information to any geographic location. The Occupational and Labor Market Information office could use it to pinpoint locations of Delaware employers, and then link any kind of data—type of industry, number of employees—to that location.

After many meetings of many committees, the two technologies—Internet and GIS—had converged in a Web site known as Career Directions. The site uses ESRI's MapObjects Internet Map Server software to bring job seekers in Delaware a wealth of information about industries, employers, and educational opportunities. The addition of data from other state agencies has turned the site into a one-stop shop for anybody looking for work in the state.

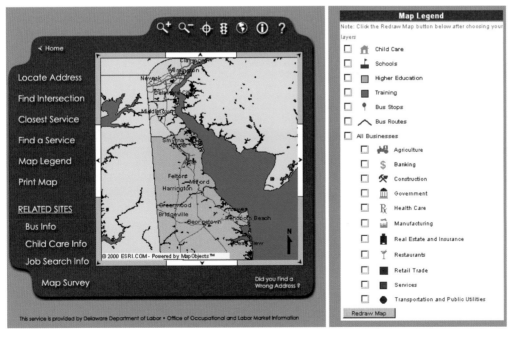

The opening screen of the Delaware Office of Occupational and Labor Market Information's Web site. The page serves as a gateway for job seekers needing information about potential employment. By checking the boxes on the right side of the frame, then hitting the "Redraw Map" button, a user brings up a variety of layers of information. Mixing and matching them customizes the map.

Laying the foundation

Among the ingredients critically important to sustained and satisfying employment is education; without it, even the most motivated and industrious job seeker won't get far. The Career Directions site, with support from the Delaware Department of Education, provides visitors with an extensive database of the state's public schools, colleges, and universities. Information on private, for-profit technical schools is also available.

The schools data, which includes contact names and telephone numbers, is accessed simply by clicking the Identify button on the interface and then clicking on the location of the school. Information about schools is updated every fall when the academic year begins.

Geographic information about public schools, of course, serves a secondary purpose: it helps a job seeker with children to make decisions about where to look for work, or even where to move the family, if that becomes necessary.

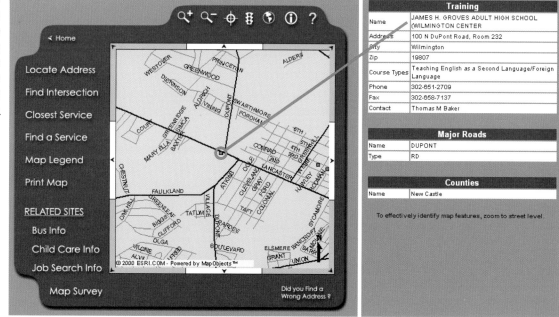

A search from the MapObjects IMS interface for "training services" in the city of Wilmington turns up many different schools, among them the James H. Groves Adult High School. Clicking the Identify button brings up additional details about Groves.

Gaining support

The availability of child care can be just as important to a job seeker as wages and working conditions—a fact employers know. The Web site's designers added child-care sites as a data layer that also can be searched.

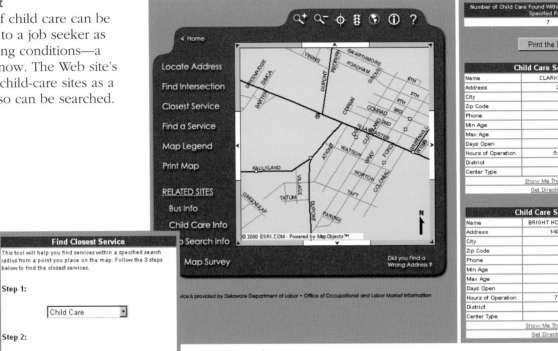

At the Career Directions site, when an employment location or a school has been found, as in the example above, its location can be used to look for related services. Drop-down lists, such as the one at the left, let you pick the service needed, at an acceptable distance. In this example, a half-mile radius drawn around the Groves Adult High School in Wilmington yields several child-care centers. Details of each child-care operation, such as the hours it's open, can then be displayed. The data on child-care facilities is compiled and kept up to date by a nonprofit agency, the Family and Workplace Connection, an example of the public–private partnership that is a critical ingredient in the Career Directions success story.

Where the jobs are

After child care, the inability to find reliable transportation is perhaps the next biggest obstacle to successful employment. The Occupational and Labor Market Information office teamed with the Delaware Administration for Regional Transit (DART) and the Department of Transportation. From the latter two agencies, the office got the data to link public transportation services to the Career Directions Web site. While linking real-time bus schedules was not yet possible, the site can nevertheless tell job seekers the next best thing: which bus lines they will need to take to get to a job or to school, and where to find the nearest bus stop for those lines. Then, a link on the page to the Web sites of transportation agencies brings up the schedules.

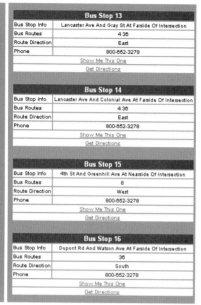

Above, a query on the Groves Adult High School brings up the location of several bus stops, marked by yellow squares. The Bus Info link brings the user to the DART Web page.

GIS to find the best route

To aid job seekers over every conceivable hurdle, the Career Directions site brings a second ESRI software technology into play. When a user asks for directions to a service such as a childcare facility or government office, ESRI's *Route*MAP™ IMS software is launched.

*Route*MAP IMS creates the correct street map, traces the path of the best route from start to finish, and even calculates both the expected time and mileage the journey should take.

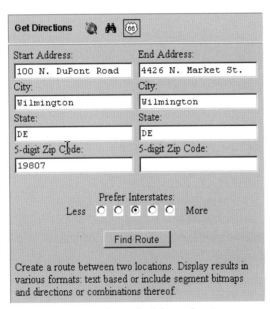

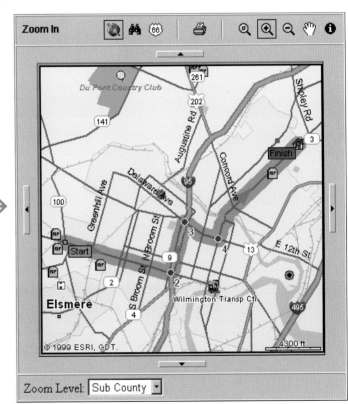

*Route*MAP IMS is launched from the Get Directions hyperlink, and brings up the dialog box above. In this example, Route*MAP has provided a map from the Groves Adult High School in Wilmington to the Occupational and Labor Market Information headquarters, one of several government offices found within a five-mile radius of the school.

Depart 100 N Dupont RD , 19807
Arrive 4426 N Market ST , 19802
Total: 4.8 mile(s) - 10 minutes

A collaborative effort

Another innovation of the Career Directions site is its operation as a modified Application Service Provider, or ASP, one of the digital economy's newest business models.

ESRI's Washington, D.C., staff and the Department of Labor created a site that integrates a voluminous amount of data. The site's servers are located and maintained at an Internet service provider that contracts with the Department of Labor for Web site hosting. The department itself retains responsibility for consolidating and maintaining data from several sources, which it then loads onto the map server and then transfers to the ISP. For example, the data for the state's child-care centers, maintained by the nonprofit Family and Workplace Connection, is updated monthly to reflect the closures and openings of new centers in that fluid business. The agency e-mails the updated data to the department, which translates it into a form that can be used by the software, and then sends it to the ISP to be placed on the servers.

ESRI continues to be involved, providing additional functionality to the Web site when requested by the department, which in turn is responding to the needs of Delaware's employers and their potential employees.

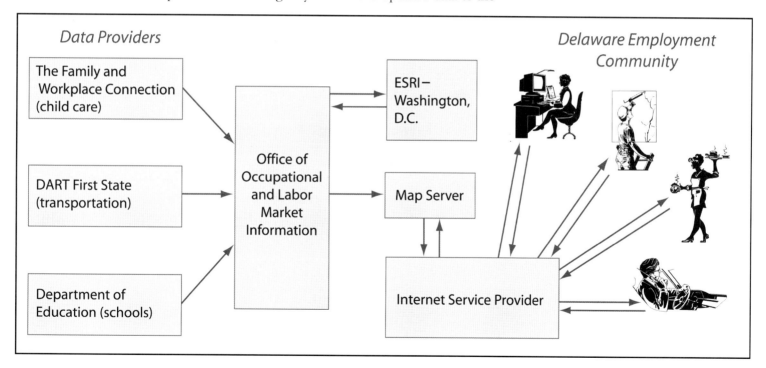

The system

The Office of Occupational and Labor Market Information uses an 866-MHz dual-processor Dell PowerEdge map server with 1 GB RAM and four 18-GB hard drives, and an Oxygen VX1 32-MB PCI video card. The Internet server is a 500-MHz Dell Dimension with 256 MB RAM and 13.6-GB hard drive. Software includes ArcView GIS 3.2, MapObjects 2.0, MapObjects IMS 2.0, *Route*MAP IMS, ArcExplorer, and Microsoft Visual Basic 6.

The URL

www.oolmi.net/Career_Directions.asp

Acknowledgments

Thanks to Lyn Anderson, Office of Occupational and Labor Market Information, Delaware Department of Labor; Gary Yakimov, formerly with the state of Delaware, and now with the Maryland Governor's Work Force Investment Board; and Rick Ayers, ESRI–Washington, D.C.

3

Northwest e-passage

While the most immediate benefit of e-government is the ability to streamline the delivery of services to the public, other, even more powerful outcomes are possible. One is the way e-government can transform the entire relationship between government and its customers, to the benefit of both. For example, the City of Tacoma used GIS maps served over the Internet as a key component of the city's economic renaissance; it found these applications did more than give better service: they gave Tacoma customers better access to better information, allowing them to make decisions for themselves at little or no additional cost to the city.

The beginning of the digital revolution turned many parts of America into boomtowns; as hardware, software, and dot-com companies prospered, so did the communities around them. In the Pacific Northwest, the fortunes of the Seattle region rose as Microsoft Corporation, based in Redmond, became the world's dominant software company.

One city initially left out of the Puget Sound high-tech boom, however, was Seattle's smaller southern neighbor, Tacoma. Private investment beginning to pour into towns to the north didn't seem to be spilling across Tacoma's border.

Public agencies stepped in. In the late 1980s and 1990s, Tacoma's downtown and nearby areas received more than $200 million in public investments. They created the Tacoma Dome; a new campus of the University of Washington; a state historical museum; a light-rail line; and a federal courthouse, renovated from a historic, long-neglected train station.

Public–private cooperation

Perhaps the most important investment in Tacoma was almost invisible, compared to the others. The local electric utility, Tacoma Power, in 1998 began laying down a 670-mile, $96-million fiber-optic network to reach every business and residence in the city. When finished in 1999, this new infrastructure, dubbed the Click! Network, spurred other telecommunications firms to make improvements in the area. The result: Tacoma became one of the most well-wired cities in the country. Not only did residents get improved TV cable services, they and the business community also gained a high-speed Internet and broadband network that rivaled, and even surpassed, the communications infrastructure available in the high-tech boomtowns to the north.

With all these public investments in place, downtown property owners began a new effort to market Tacoma to businesses looking for a Northwest connection. Not only did Tacoma have all its new public improvements to offer, it could also boast a lower cost of doing business and cost of living than Seattle could, even though it was just as close to the Northwest's natural beauty.

To ensure they had an accurate, up-to-date picture of property and office space available for the new business they hoped to attract, the city's Economic Development Department switched to a new system based on ArcView GIS.

Public investment in Tacoma helped it blossom. The University of Washington opened a new Tacoma campus, left, and federal funds turned the dilapidated but historic downtown railway terminus, below, into a new federal courthouse.

From paper to digital

ArcView GIS simplifies the work of those in a city, such as economic development analysts, who need large amounts of current information about a location, such as a business district. In Tacoma, analysts had to thumb through a thick binder of downtown property statistics that was updated only every two years. With ArcView GIS installed on the department's desktops, analysts could get the same information simply by clicking a mouse on an ArcView GIS digital map of the city. They would immediately see the information the city had on record about a piece of property: size, ownership, property taxes paid, zoning, the kinds of permits that had been issued there, and much more. Moreover, that information could be updated more quickly and easily, in tune with a dynamic property market.

ArcView GIS made information more accessible to analysts, who then made it available to the public. The next question was: why not make that same information available to the public directly, without the city as the middleman?

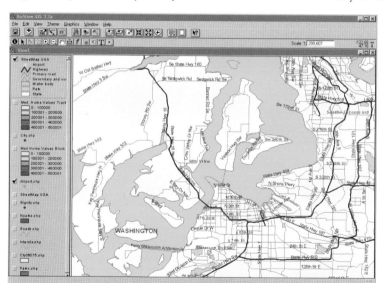 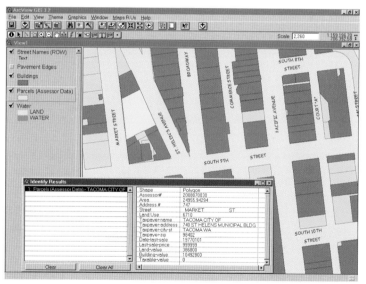

Tacoma as seen through ArcView GIS. The view of the Puget Sound area, left, uses the ArcView GIS StreetMap™ extension; the other is a parcel map used by city employees.

Maps on the Internet

The answer was that there was no reason not to—especially given the ease with which it could be done, using the Internet Map Server (IMS) extension to ESRI's ArcView GIS software.

The extension allows anyone creating maps in ArcView GIS to deliver those maps over the Internet to anyone with a standard browser. More importantly, simple customization can make those maps interactive, letting you narrow the thematic focus of a map, or to pose questions about the features in it.

To get the Web site up and running required a partnership among Tacoma's GIS and Economic Development departments, the Local Development Council (LDC), and the government of Pierce County. The city agreed to supply the software and hardware and GIS expertise, and to host the Web site on its servers. The LDC, which was made up of commercial property owners, would keep the database updated with current information about properties and related data.

The new site, dubbed TacomaSpace, immediately gave the city a formidable Web presence, moreso than the run-of-the-mill electronic brochures used by many cities. TacomaSpace allows developers and prospective businesses a tool to gauge for themselves the real economic potential in Tacoma.

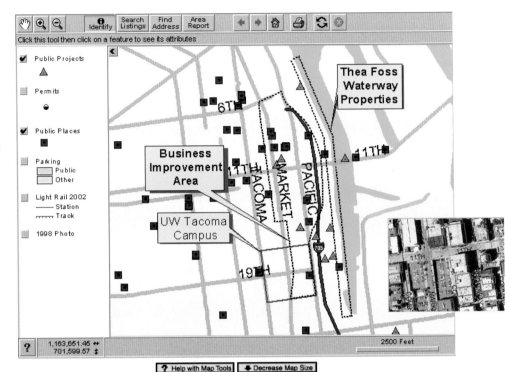

A simple map of downtown Tacoma from the TacomaSpace home page. In the left margin are the themes, containing common features, which can be turned on or off in the map itself by checking the small boxes. Navigation buttons at the top of the map allow the user to zoom in and out, to move the map, and to bring up additional information in report form. The inset shows a portion of the "1998 photo" theme, which is an extremely detailed aerial photo of the city. Properties available for lease or sale are outlined in purple in the photo.

Through Tacoma by Web

The opening page of the site gives you a choice between accessing a map of the downtown business core, or a map of one of a dozen business districts throughout the city. Buttons on each allow map navigation and identification of features.

The three buttons of greatest interest to the real-estate and business community are labeled Search Listings, Find Address, and Area Report.

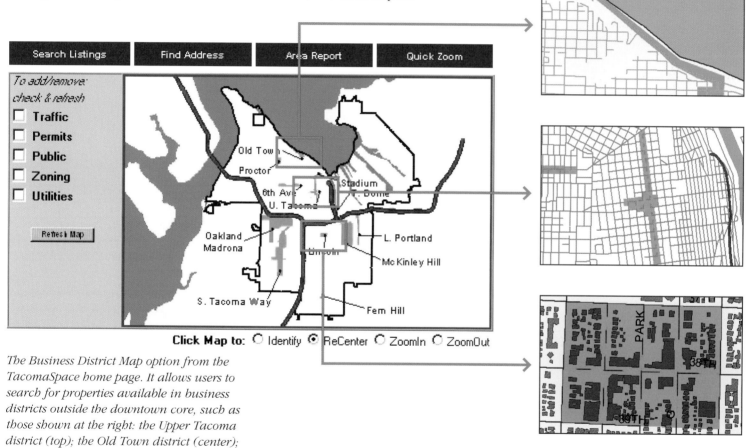

The Business District Map option from the TacomaSpace home page. It allows users to search for properties available in business districts outside the downtown core, such as those shown at the right: the Upper Tacoma district (top); the Old Town district (center); and the Lincoln district (bottom).

Finding the perfect property

The Area Report function allows you to find out what kinds of economic activity are occurring around a site, as well as information about nearby demographics, parking, and any incentives local government may be offering in that area.

The Find Address button lets you enter an address; ArcView IMS then zooms the map to that spot. Additional information is then available: whether space is available in a building, when its lease is up, the year it was built, even photos.

The Search Listings button lets you access all of the properties that the local real-estate community currently has available.

With this function, you specify criteria for property you're interested in: its size, a lease or sale, whether what you want is commercial, retail, office, industrial, or vacant land.

Once the criteria are set, the site comes up with a list of properties that fit them. Clicking on the address of one of those properties then brings up a new screen with a new, detailed locator map.

It also brings up even more details: exact square footage, restrictions on development, and utility services.

ESRI's ArcView Business Analyst software, linked to this function, lets you research demographics and business statistics about the area.

TacomaSpace also lets you jump from the property information graphic to the Web sites of other related organizations, such as the Pierce County Assessor's Office.

The Search Listing function alone sets TacomaSpace apart from most other real-estate sites—most present lists of properties to you, and it's up to you to find the time to rummage through them for something that fits your needs.

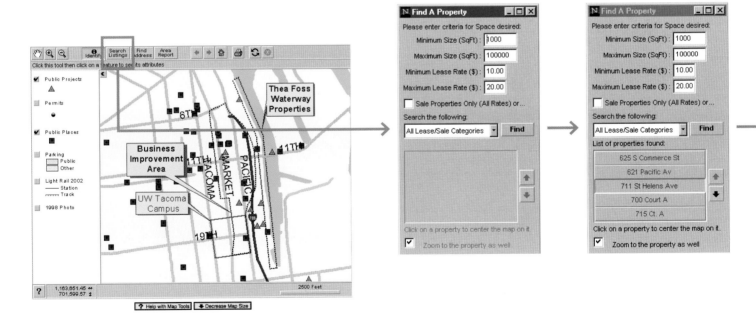

Overcoming doubts

Commercial real-estate brokers were at first concerned about whether they would be shortchanging themselves by freely giving out information on the Web site that many considered proprietary. But they quickly realized that most of the information was in the public record anyway, and that TacomaSpace was helping give that information wider distribution. The site did not change their identity as brokers; any sales would still have to go through them.

Downtown Tacoma is now booming, with high-tech firms with names such as Amazon.com, InsynQ, and Alcatel setting up shop, their executives boasting about the low lease rates and the great views of Puget Sound and Mt. Rainier. The Click! Network, the public improvements, the marketing campaign, and TacomaSpace have all contributed to the boom.

TacomaSpace has changed the way the city's Economic Development Department goes about its business. Now, prospective tenants and owners can search through property listings on their own, browsing property online in a way they never could before.

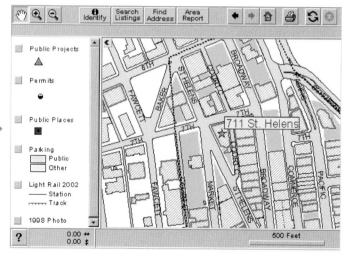

The system

300-MHz PC with 128 MB RAM and 4-GB hard drive, for both map and Web server. Individual property data is maintained in a Microsoft Access database connected to ArcView GIS. The property database is updated on a separate PC and then copied to the map/ Web server.

The URL

www.wiredcityusa.com

Acknowledgments

Thanks to Merten Bangemann–Johnson, Mike Murnane, and Bill Bogue of the City of Tacoma, and to Diane Lachel and Lee Root of Tacoma Power.

OzarkIMS

4

"Next-generation technology" is a phrase that tends to evoke a picture of huge, new innovations, created by behemothic research outfits with billion-dollar budgets. But when it comes to the newest generation technology for serving maps over the Internet, you don't need the latest climate-controlled Cray supercomputer. Among the first users of this technology, ArcIMS, is a small city nestled in the quiet hills of northwest Arkansas, whose roots reach back to the early nineteenth century. The city of Fayetteville brought its first GIS mapping services to the Web using ArcIMS because it was technology that could be used right out of the box and could also keep pace as Fayetteville's e-government services grew. Even, if necessary, to behemothic proportions.

After reading in a 1998 book, *Serving Maps on the Internet,* how ESRI's Internet Map Server technology could bring customer service and public visibility to new heights, the GIS team in Fayetteville figured IMS was a technology that could clearly benefit the city. They had years of investment in data for internal use that they wanted to see put to use by the public. But rather than construct a GIS Web site from different pieces of different applications, they searched for a complete technology solution they could use as a solid foundation for future growth that would also be easy for the public to use. ArcIMS fit those requirements. Because of its potential, the Fayetteville GIS team held off putting any map services on the city Web pages until the 2000 release of ArcIMS 3.

ArcFayetteville

With ArcIMS, you can create complete and self-contained e-government mapping services on a Web site that incorporate the kind of GIS power available inside an organization. Its range is broad: it can build a complex site that lets your visitors do such tasks as measure distances from one end of town to the other, or find addresses for them, or create buffers around locations. But it can also let you build GIS maps that are easy to understand and simple to navigate, so you don't scare away new visitors to your Internet site.

ArcIMS is highly scalable, meaning that if there is a sudden surge of users to your site, it will keep up with the demand, instead of locking up your servers and turning visitors away with one of those annoying "Sorry, server busy" messages.

All of these features of ArcIMS, plus its ease of use out of the box, made it Fayetteville's solution of choice. The Fayetteville GIS team was looking for speed, so it could get its interactive maps up and running in time for the November 2000 elections.

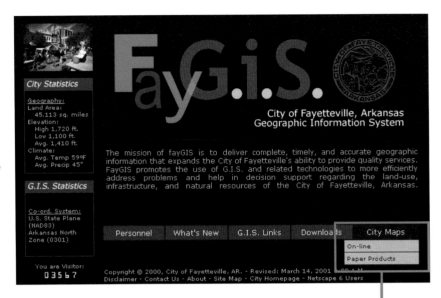

The Fayetteville ArcIMS introductory interface is highly customized; its menu choices drop down automatically when the cursor rolls over them. The City Statistics and GIS Statistics boxes, above, contain several different informational sets on a continuous loop, allowing more reference facts about the city to be displayed than would appear on a static display.

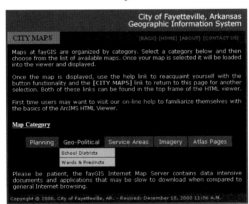

Look all ways before you vote

That elections application, known as the FayGIS Wards and Precincts page, answers election questions with maps that show the locations of election jurisdictions and polling places.

A hyperlink function built into ArcIMS lets a voter bring up specific information about one of these features—such as the address of a polling place—merely by clicking on the feature on the map.

Four wards meet in this area of the city. The lightning bolt icon in the Themes list signifies an additional hyperlink function that in this case brings up photos of council members.

Illuminating the twilight zone

Zoning regulations can be among the most confusing of subjects for citizens: not only can allowable uses change from one side of the street to the other, myriad subclauses and exceptions to zoning use further complicate matters.

The Fayetteville GIS team sheds some light on the subject with a detailed interactive zoning map on its Web site.

Combined with a map of all the building shapes—known as "footprints"—in the city, questions about exactly where something can be built and where it can't is one no longer shrouded in mystery.

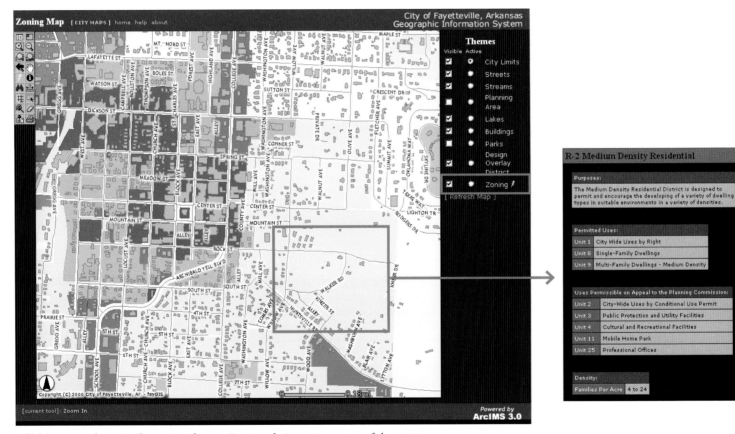

Clicking anywhere in the active theme, Zoning, brings up a copy of the exact language of the relevant code, giving Fayetteville residents a clear picture of what buildings are allowed where.

Getting back to the basics

One of the best things about an interactive GIS on the Web is that it can help residents get answers with relative ease to basic questions about common government functions. Especially for newcomers, finding out about trash pick-up or street-sweeping schedules, or what schools their kids will attend,

can involve hours thumbing through phone books and being put on hold.

The FayGIS School District maps typify how ArcIMS can ease this process for residents. Detailed maps of district boundaries, school attendance zones, and addresses help eliminate the common aggravations too often associated with government.

FayGIS maps of school districts, with street-level detail, below. The list of layers to the right of the large map shows which themes you can display. You can choose between that or the legend of the small map.

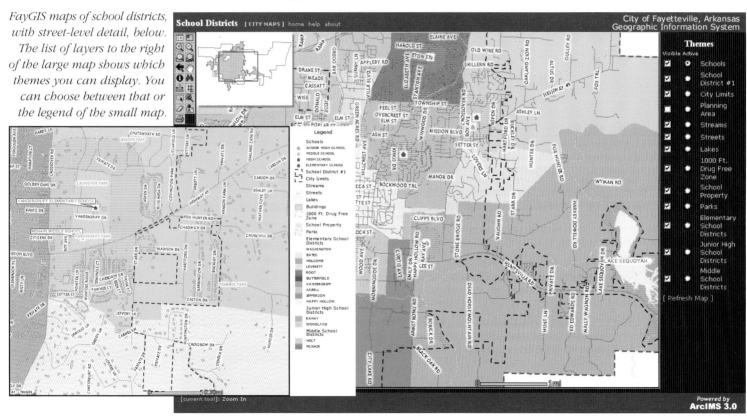

Station to station

Government does more than regulate, of course—protection of people and property is fundamental. The Fire Service Areas page shows residents just how some of these protection responsibilities have been allocated throughout the city.

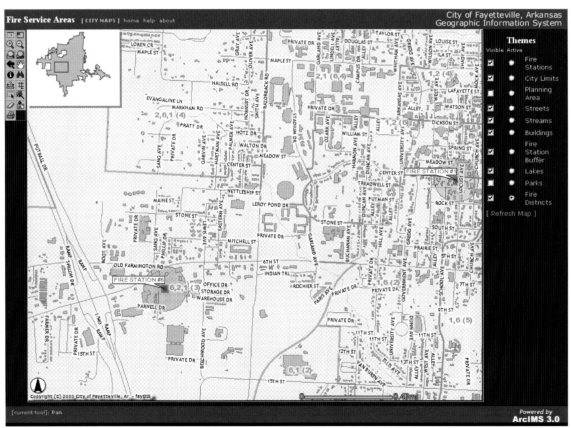

The red lines delineate each fire station's primary response area. The numbers in blue show the order in which stations will respond to an emergency in an area; the number in parentheses at the end is the station responsible for backing up the primary station.

Mapping history

While interactive mapping is usually thought of as a utility, the FayGIS site is unique in acknowledging and showing that Fayetteville is made up of more than just administrative districts and functions, and that it also has a history. ·

The Fayetteville Annexations page is GIS in a different context, showing more than present reality, but also how the city grew from a small, square-shaped town in 1870 to the vibrant city that it is today.

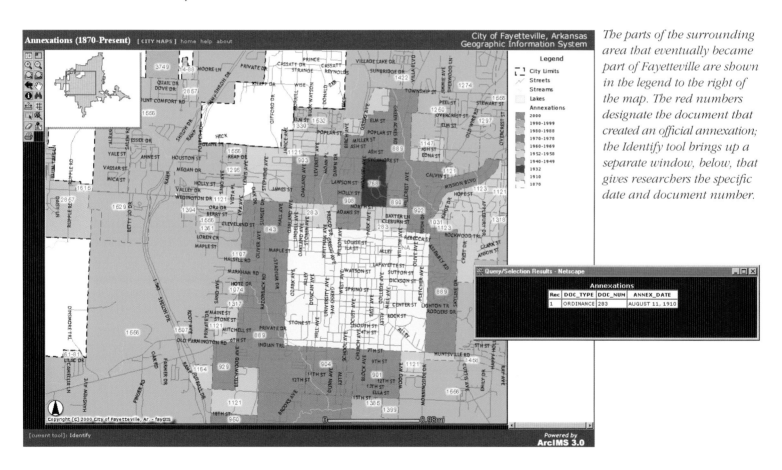

The parts of the surrounding area that eventually became part of Fayetteville are shown in the legend to the right of the map. The red numbers designate the document that created an official annexation; the Identify tool brings up a separate window, below, that gives researchers the specific date and document number.

The system

800-MHz IBM Netfinity 3500 M20 with 1 GB RAM and three 18.2-GB hard drives in a RAID-5 system, using ArcIMS 3, Microsoft Windows NT Server SP5, Microsoft Internet Information Server 4.0, and Servlet Exec 3.0.

The URL

www.faygis.org

Acknowledgments

Thanks to Scott Caldwell, City of Fayetteville.

5

Property, parcels, and plots

Keeping track of who owns what piece of property, parcel, and plot of land in an area is the focus of much local government activity, not surprising considering that the bulk of their revenues come from taxes on those pieces of property, parcels, and plots. As they turn increasingly to e-government solutions to streamline their operations, increasing numbers of local governments are putting property-related applications on their e-government Web sites. Few are doing so as intensely as Greene County, Ohio, where the county Auditor's Office has put virtually all of its property-related functions onto its Web site, using GIS technology.

Ohio is a place where marketing firms often come to try out new products because it serves as such a good barometer of Americana, and Greene County is in many ways typically Ohioan: a mixture of rural and suburban populations totaling about 150,000, with some colleges located within its boundaries, as well as a major military facility, Wright–Patterson Air Force Base.

Being such an emblematic community means that what they have been able to do with GIS in Greene County could probably be done in many other towns across the country.

GIS from way back when

Admittedly, Greene County has had something of a head start. It began pursuing GIS programs back in the high-tech equivalent of the Ice Age: the early 1990s. Inspired by what another Ohio county was doing with GIS technology, the Greene County auditor put together a $60,000 pilot project to scan all of the county's property tax maps into digital format.

The results of this experiment demonstrated to county officials, many of whom didn't know much about the technology at all, just how useful GIS could be—that it could handle large amounts of complex data like property records, that it could do it quickly, and that it could integrate the work of several different departments.

Based on that experience, the county authorized an aerial survey of the entire county, and thereafter, the conversion of the entire Greene County parcel database into a GIS, using ESRI's ARC/INFO software. This complex and painstaking task, which required matching the exact legally defined geographic coordinates of every piece of property in the county, as well as street dimensions and administrative boundary lines, took three years.

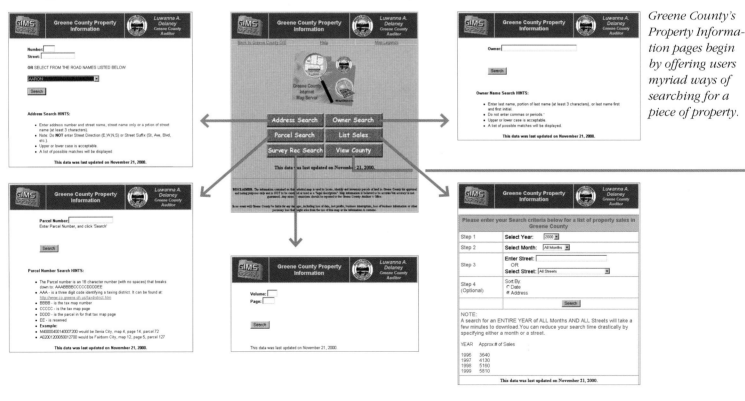

Greene County's Property Information pages begin by offering users myriad ways of searching for a piece of property.

All roads lead to property

County officials figured that building such a comprehensive and accurate database, although time-consuming, would ultimately result in a better GIS system that would ultimately benefit everyone. A good foundation would allow for more flexibility down the road, and would naturally prevent the need for expensive retrofitting when new applications were needed. That would advance the county's ultimate goal of making the system as useful as possible, because many other county agencies would be able to use it. That in turn would make it as useful to as much of the public as possible.

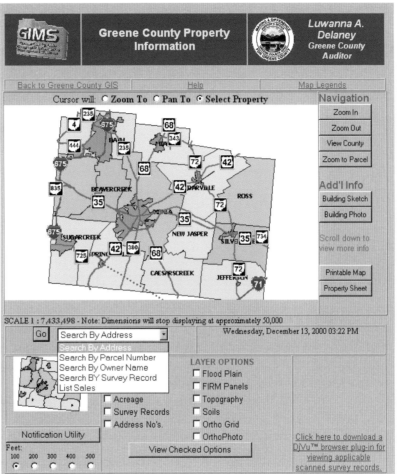

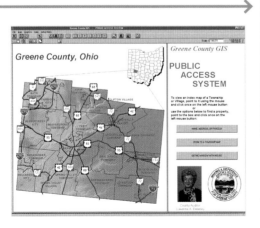

The simple appearance of the Greene County map server page, above, belies robust functionality underlying it. Built on MapObjects IMS technology, it is linked to voluminous amounts of data. A similar application, left, the Public Access System, is used in Greene County offices and, even though it has a similar appearance, is built on ArcView GIS.

Tying it all together

A GIS works by linking a digital representation of a geographic location to a database of information about that location. That is just how Greene County's system works: all the information about a parcel is linked to a digital map. Greene County's achievement is that it has linked virtually every piece of information available, from the most obvious to the most arcane. It has then gone one step farther and made all that information available through the Internet, and at its public-access sites. Any property owner's records can be found; although some agencies hide IMS information about property owners, or restrict information about certain people, Ohio's open records laws are among the most liberal in the nation, and so property information is available about any owner.

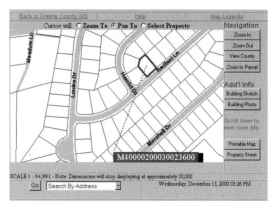

PROPERTY INFORMATION FOR: M40000200030023600			
DISTRICT:	XENIA CITY	BUILDING SQFT:	2107
OWNER NAME:	DELANEY FLOYD E	BEDROOMS:	3
OWNER (cont.):	& L A	FAMILY ROOMS:	1
ADDRESS:	500 REDBUD LA	FULL BATHS:	2
ACRES:		HALF BATHS:	0
YRBLT:	1959	BASEMENT:	NONE
STYLE:	RANCH	FIREPLACE:	1
LEGAL1:	STADIUM PARK 5 ALL	CONSTRUCTION:	BRICK
LEGAL2:	LOT 105	CLASS:	
LEGAL3:	500 RED BUD LANE	UTILITIES:	ALL PUBLIC
MAIL ADDRESS:	500 REDBUD LA	ADD'L UTILITIES:	
MAIL ADDRESS:	XENIA OH 45385	UTILITY CONT'D:	
APPRAISED LAND VALUES:	20270	HEATING:	CENTRAL AIR CONDITION
APPRAISED BUILDING:	99000	HEAT FUEL:	ELECTRIC
APPRAISED TOTAL:	119270	SALES DATE:	
ASSESSED LAND VALUES:	7090	SALES PRICE:	
ASSESSED BUILDING:	34650	CONVEYANCE NO.:	
ASSESSED TOTAL:	41740	ACT FRONTAGE:	82
TOTAL TAXES:	1568.3	EFF FRONTAGE:	82
TAXES DUE:	0	DEPTH:	97

DEED INFORMATION FOR: M40000200030023600					
VOLUME	PAGE	DATE	INST TYPE	PARTY	RELATIONSHIP
0980	0288	3/11/96	DEED	DELANEY LUWANNA A	Grantee
0980	0288	3/11/96	DEED	DELANEY LUWANNA A	Grantor
0980	0288	3/11/96	DEED	DELANEY FLOYD EUGENE	Grantee
0980	0288	3/11/96	DEED	DELANEY FLOYD EUGENE	Grantor
0334	0067	11/6/61		DELANEY FLOYD E	Grantee

Left, the basic map that appears in the MapObjects IMS application when a particular parcel is selected, in this case, the home of Greene County Auditor Luwanna Delaney and her husband. Above, the detailed property information that appears on the same screen below the map.

You can leave your pen at home

What Greene County saves its residents, of course, is time—time spent driving to a county office to plow through microfiches for the right property maps and the information found there; time spent waiting at the counter for the clerk to find the map you want; time spent looking for the pen you swore you had just a moment ago; time spent trudging over to a different Greene County office that has the information you thought was at the first office; time spent driving home, wishing there was a better way.

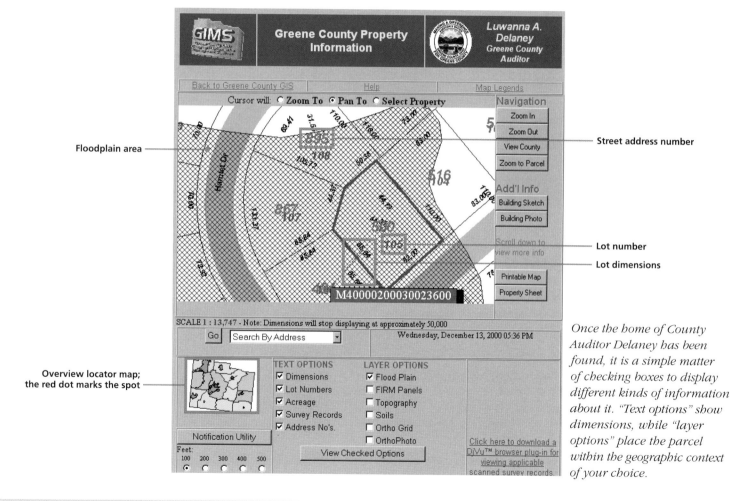

Floodplain area

Street address number

Lot number

Lot dimensions

Overview locator map; the red dot marks the spot

Once the home of County Auditor Delaney has been found, it is a simple matter of checking boxes to display different kinds of information about it. "Text options" show dimensions, while "layer options" place the parcel within the geographic context of your choice.

One-stop access

While the Greene County system had its genesis in the auditor's office, information that other departments maintain is being integrated into the system, extending GIS efficiencies to other parts of the enterprise.

For example, the county integrated a document-imaging system into the GIS. This means a parcel is linked not just to primary information about the parcel, but also to specific legal documents. These include photos, schematic drawings, survey records, deeds, and other official documents held by the recorder's office. Scanned, placed in a database, then linked, they are available for instantaneous retrieval.

Among the painstaking chores done when Greene County converted its property tax information to digital format for use in the GIS was matching property drawings to the aerial photos. Known as digitizing, it is a process both laborious and time-consuming because each line from the photo must be traced and transferred to the map. But the county's early adoption of GIS technology made this task less onerous because it already had the equipment and software in place to do the work on a PC instead of manually. The results are shown at the right: when you click the OrthoPhoto box, the aerial photo is displayed, and other information is overlaid. Ortho grid numbers identify photos that you can buy for a particular area.

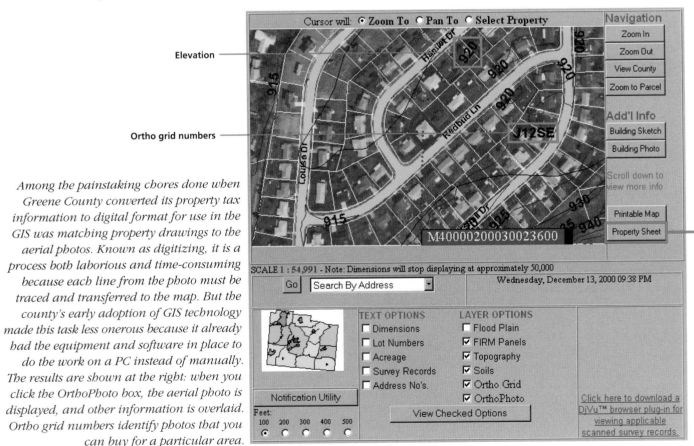

Easing assessment

Having detailed and in-depth records about a piece of property allows assessors in the auditor's office to cut down on the amount of preparation needed for a property analysis. While it doesn't obviate the need to make physical inspections when they are warranted, the availability of so much information about the property via the GIS makes the whole process more efficient.

It also helps with customer service. Property owners often phone the auditor's office with questions about an assessment. Many of their questions can be answered immediately because so much information about the property is available so quickly to the person taking the call.

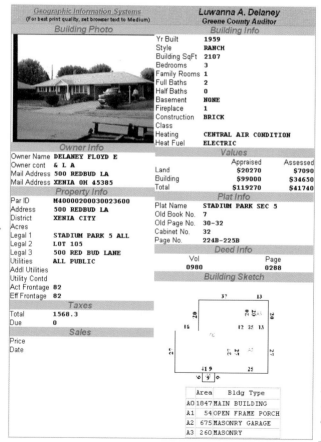

Geographic Information Systems
(For best print quality, set browser text to Medium)

Luwanna A. Delaney
Greene County Auditor

Building Photo

Building Info

Yr Built	1959
Style	RANCH
Building SqFt	2107
Bedrooms	3
Family Rooms	1
Full Baths	2
Half Baths	0
Basement	NONE
Fireplace	1
Construction Class	BRICK
Heating	CENTRAL AIR CONDITION
Heat Fuel	ELECTRIC

Owner Info

Owner Name	DELANEY FLOYD E
Owner cont	& L A
Mail Address	500 REDBUD LA
Mail Address	XENIA OH 45385

Values

	Appraised	Assessed
Land	$20270	$7090
Building	$99000	$34650
Total	$119270	$41740

Property Info

Par ID	M40000200030023600
Address	500 REDBUD LA
District	XENIA CITY
Acres	
Legal 1	STADIUM PARK 5 ALL
Legal 2	LOT 105
Legal 3	500 RED BUD LANE
Utilities	ALL PUBLIC
Addl Utilities	
Utility Contd	
Act Frontage	82
Eff Frontage	82

Plat Info

Plat Name	STADIUM PARK SEC 5
Old Book No.	7
Old Page No.	30-32
Cabinet No.	32
Page No.	224B-225B

Deed Info

Vol	Page
0980	0288

Taxes

Total	1568.3
Due	0

Sales

Price	
Date	

Building Sketch

Area		Bldg Type
AO	1847	MAIN BUILDING
A1	54	OPEN FRAME PORCH
A2	675	MASONRY GARAGE
A3	260	MASONRY

Pressing the Building Photo or Building Sketch button on the property sheet encapsulates information available from other portions of the Web site.

The system

Hardware: Gateway 7400 dual 800-MHz server with 1 GB RAM for SDE data, and a dual 300-MHz server with 512 MB RAM for IMS.

Software: ArcInfo™ 8.02, ArcView GIS 3.2a, MapObjects IMS 2.0, ArcSDE™ 8.02 for Microsoft SQL Server, Oracle 7.3, Microsoft Access 2000, LizardTech MrSID™ Geospatial Encoder 1.4, LizardTech DjVu Encoder 3.1.

The URL

www.co.greene.oh.us

Acknowledgments

Thanks to Steve Tomcisin, GIS director, Greene County Auditor's Office, and to Luwanna Delaney, Greene County auditor.

6

E-votes, e scores

Arizonans pride themselves on their independence, on a certain disinclination to follow the crowd. Innovation often springs from such an ethos, and in the realm of e-government, innovation has helped Arizona become recognized as a leader. In 1999 the state's online system for automobile registration, known as ServiceArizona, was awarded a top prize for innovation by the National Association of State Information Resource Executives. And in early 2000, Arizona held the first election in the country in which voters cast ballots over the Internet. In Arizona's rural Yavapai County, GIS is being used online to aid the democratic process in another way.

Yavapai County, located north and west of Phoenix, is huge; you could fit any of a number of East Coast states inside it and have space left over. Its eight-thousand-plus square miles contain vast stretches of desert, scrub land, and forest. Home to some of the most striking desert landscape imaginable, Yavapai County affords its 155,000 residents plenty of room.

When you have lots of land, you have lots of land to keep track of. Yavapai County employees in the county seat of Prescott noticed that residents—their customers—seemed to have a higher-than-average interest in land and property issues: the demand for county maps was high. One likely reason is the number of people who have chosen Yavapai as their place of retirement. Being retired, they had both the time and the inclination to keep close tabs on the world around them. The result was they kept the Yavapai GIS department jumping, making maps.

A multitude of jurisdictions

In addition, Yavapai is a complex area administratively. Other jurisdictions, including federal and state agencies, own sizable chunks of land within the county, which are themselves criss-crossed by a variety of legislative and political jurisdictions.

Residents trying to identify where their property taxes were going, or which elected official they should complain to, or which elected official they wanted to throw out of office, found that doing so could be quite a chore. Filling their requests for electoral information and maps kept county employees running for the phones or the front counter, sometimes to the detriment of other responsibilities.

Moreover, residents who did make it to the front counter to pick up a map often faced a different problem, Yavapai's sheer size. Simply getting to the county seat of Prescott could take half a day over sinuous mountain and desert roads.

Using a GIS over the Internet seemed like a natural solution to these problems. Anyone with a telephone line could get to the county Web page. And once at the county Web page, anyone could access the power of GIS—to pick and choose the layers of information they wanted, then combine them to create the map they needed.

And having residents themselves create these maps would free up valuable, taxpayer-funded county time.

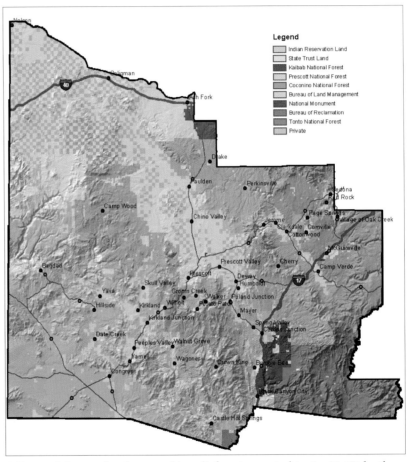

A Yavapai County land ownership relief map, created in ArcMap™ by the county GIS department, shows the variety of jurisdictions and geologies in the eight-thousand-square-mile area.

The importance of being property

However, another group of map devotees was also demanding time and attention—commercial users. Appraisers, real-estate brokers, title companies, and even graphics firms—who needed exact lot dimensions for blueprints—all clamored for information from the GIS department as well. So Yavapai's first GIS offering over the Internet in late 1998 was a property-map application built on ESRI's MapObjects Internet Map Server technology.

It was a success almost overnight, providing customers with what they needed, while relieving pressures on the department.

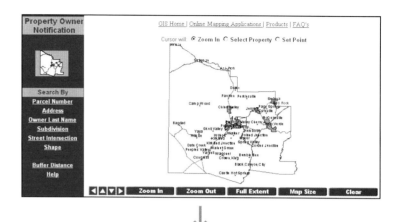

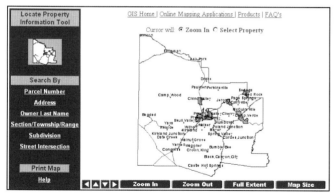

The first Internet GIS tool the county put up targeted property applications, helping users to locate and identify individual parcels.

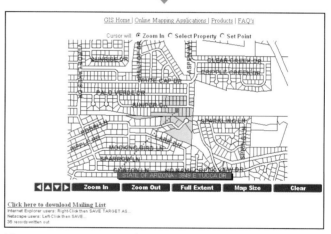

The Property Owner Notification application, above, lets users pinpoint a piece of property—here, one owned by the state of Arizona—and then select parcels within a specified distance for notification about proposed administrative processes—zoning changes, for example. The hyperlink at the bottom left gives users a mailing list of the names and addresses of property owners within that specified distance who need to be notified.

Mapping the vote

Bouyed by its property-map success, the GIS department then looked for other problems to solve—such as helping ordinary folks figure out which of the county's myriad jurisdictions they lived and voted in. This task was complicated by a fundamental aspect of democracy, the proportional representation form of government, which dictates that a governing body be made up of districts roughly equal in population.

In a sprawling area like Yavapai County, where population groups are so scattered, this can make for some odd-shaped districts that rarely line up neatly. You could be in one electoral district and your neighbor across the street could be in a different one.

But a GIS works by allowing the user to place layers of similar geographic features—such as rivers or electoral districts—on top of each other, so that each can be seen in relationship to the others. It was on this idea that the county's Election Map Tool was based.

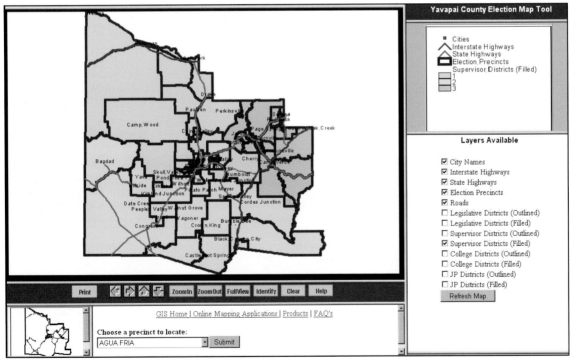

The default opening screen of the Yavapai County Election Map Tool shows the user a basic map of the county and its division into three supervisorial districts. Users can pick and choose the boundaries they want to see from the list in the right frame. At the bottom is a tool for highlighting an individual voting precinct. The larger map within the context of the entire county appears at the lower left.

Integrating jurisdictions

The Election Map Tool integrates several different Arizona electoral districts as layers. By picking and choosing among layers, residents can build their own maps to find their geographic place within supervisorial, judicial, legislative, or college districts.

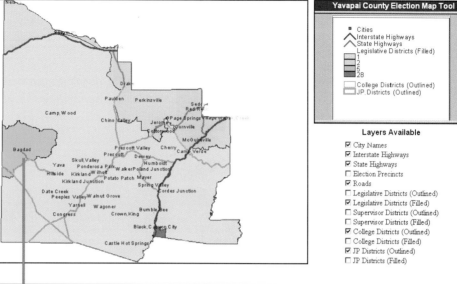

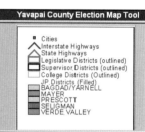

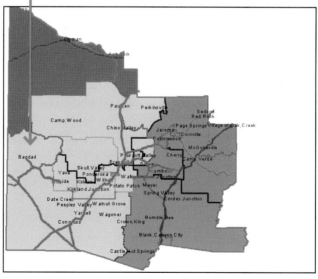

The Layers Available box allows users to view districts either as solid units or as outlines. The optimal way to use this tool is to choose one layer to view as a solid, allowing the others to be viewed in outline; choosing to view the solid districts also brings up their names in the legend (although names of districts can always be seen using the Identify tool). In the example above, state assembly districts in the county are viewed as solids, while the others are outlines. The example at the right shows solid judicial districts, while the assembly districts are outlines. The arrow shows that the solid 5th district in the top map is still identifiable in the bottom one.

Finding the ballot box

But all this electoral information won't do much good if people don't vote. Among the most common excuses for not voting is that people don't know what precinct they're in, or where to go to vote. Yavapai residents who call up the Electoral Map Tools site can't use such excuses.

First of all, the home page, by default, is already divided into election precincts. If that's not enough, a drop-down box offers the names of all the precincts in the county. When the user chooses one, the map tool then zooms to a close-up of the district so detailed that it includes individual streets.

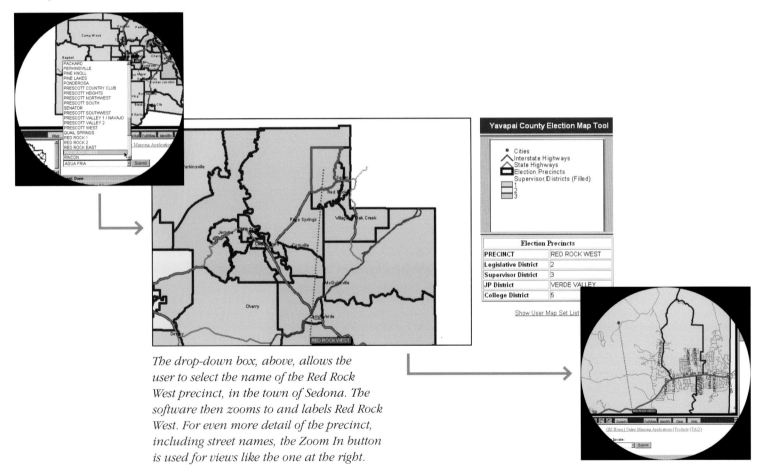

The drop-down box, above, allows the user to select the name of the Red Rock West precinct, in the town of Sedona. The software then zooms to and labels Red Rock West. For even more detail of the precinct, including street names, the Zoom In button is used for views like the one at the right.

Integrating applications

The GIS department's IMS construction work was made easier through templates—standardized formats—supplied by ESRI with MapObjects IMS. The property map application used early versions of them, while the Election Map Tool used those in general release.

Those templates can easily be customized, and were, in Yavapai's applications. For example, the GIS department wanted to simplify the appearance and functionality of the property map application, and modified the template so that the browser—the window that people see and which contains the maps—didn't use frames. Such frames aren't necessary for a Web application that has only one purpose, and they can confuse inexperienced users.

But for the Election Map Tool, the department wanted users to be able to choose from different layers, a task made easier with the frame-enabling capabilities that come with the IMS templates.

Yavapai's Internet map services have proved out the original expectation that they would fulfill a need of this unique land and community: of the twelve thousand hits the Web site gets in a month, ten thousand are for GIS map services.

Which shows that in Yavapai, they are paying attention. And thanks to GIS, that attention gets satisfied—without undue stress on taxpayer-funded resources.

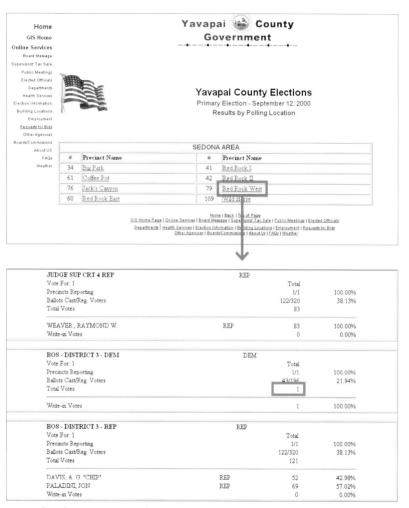

From the Election Map Tool page, community activists and others who pay close attention to elections can jump to the county's elections department, where detailed results and other information can be found. Here, the Red Rock West precinct evinces little enthusiasm for the Democratic party.

The system

Yavapai runs its IMS applications on a 266-MHz dual-processor PC with 256 MB RAM and 9-GB hard drive. Windows NT Server 4.0 is the operating system.

The URL

www.co.yavapai.az.us

Acknowledgments

Thanks to Scott Edwards, Kris Estes, Kevin Blake, and John Thomas of Yavapai County GIS, and to Jim Holst, Yavapai County administrator.

Extending the Pulaski

7

The Cerro Grande burn in the spring may have been an omen that the fire season of 2000 was not going to be normal. Cerro Grande began peacefully enough as a controlled burn, but quickly overwhelmed its handlers to become New Mexico's worst fire disaster ever. Much worse was to come. By late summer, there was conflagration across the western United States and by late autumn when it was all over, almost seven million acres had burned, more than twice the annual average—destruction on a phenomenal scale. A disaster of such magnitude compelled fire managers to look everywhere for help, to think in new ways. From this need came a nation-sized wildfire GIS system that provided interactive maps on the World Wide Web to fire managers in any location—letting them, and the public, see not just where every fire was burning in the country, but also its size and its potential for destruction. With GIS, they could make strategic decisions about deploying fire-fighting resources in a way they had never been able to before.

Technology's march sometimes seems inexorable, but there are some areas where computers and software have yet to make much of a dent. Fighting forest fires, for example, doesn't require a big bag of subtle technological tricks. To be sure, there are lots of vehicles and lots of aircraft, though many are of a decidedly low-tech, DC-7 vintage. But

on the ground, the tools used for fighting a fire are about as fundamental as fire itself. One of the most basic is the Pulaski, a short-handled combination axe and hoe, used for cutting through brush and earth to create fire lines. It's a stout, brutal tool, exactly what's needed to battle a stout, brutal wildfire.

Fighting fire with paper

But low-tech has its limitations. Maps, for instance, are essential for assessing the geography around a wildfire, and for tactical planning. Most firefighters still use paper ones effectively.

They also use tape and scissors—to cut and paste these government-issue topographic paper maps together to make one that fits the actual area in which they're working. In 2000, some fires were so big they encompassed 120 such maps.

Fire raging all around them, elk take refuge in a fork of Montana's Bitterroot River in August 2000.

John McColgan/Alaska Fire Service, Bureau of Land Management

Out of crisis, innovation

Battling the fires of summer 2000 required unprecedented numbers of firefighters and tons of equipment. With both in short supply, national and regional firefighting managers had to pull off a monumental juggling act, moving supplies and personnel from one end of the country to another, often on an hour's notice. Among the myriad agencies responsible for this decision making was the Great Basin Multi-Agency Coordination (MAC) Group, which by early July was meeting twice a day—to assess the rapidly changing needs of incident commanders, and then moving always-insufficient resources where they were needed.

It was the toughest kind of decision making, not made any easier by the fact that much of the incoming information was in the form of written reports full of words and numbers and tables: as copy machines whirred endlessly, wildfires were destroying forests and homes and lives.

In late July, one of the MAC Group's members, veteran firefighter Robert Plantrich of the Bureau of Indian Affairs, sent off a memo urging the deployment of a new kind of tool, GIS, to help with some of the critical decision making.

The response was astonishingly swift for a bureaucracy. Within hours, a conference call among GIS-savvy agencies was taking place, laying the groundwork for a national wildfire Web site.

Finding and allocating resources such as hand crews, planes, and vehicles was one of the toughest jobs of fire managers during the summer of 2000.

One-stop fire shop

Within days of the conference call, the Geospatial Multi-Agency Coordination Group (GeoMAC) Web site was under construction on servers at the USGS Rocky Mountain Mapping Center in Colorado. The site would be a one-stop information shop, where fire-fighting managers could go for the latest information on fires. Many agencies had a voice in the site's development, notably the Forest Service and the Bureau of Land Management, on whose land many wildfires were burning. That meant there was a wealth of data that site developers from ESRI and the USGS could work with. That wealth, in turn, dictated the technology that would be used—ArcIMS, the most up-to-date software for serving maps over the Internet, and ArcSDE for Oracle8i™ to harness all that data, with IBM providing a server.

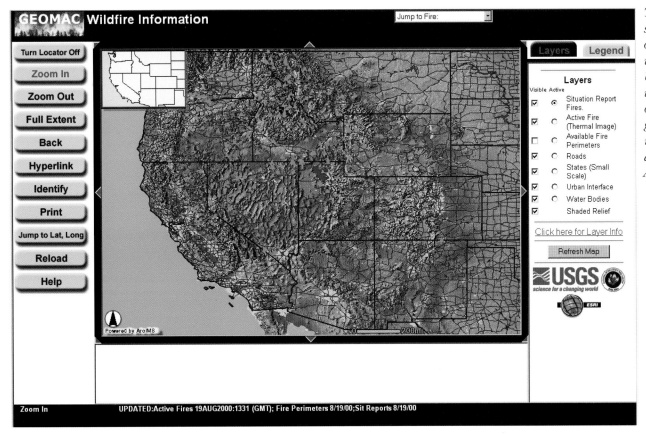

The GeoMAC Web site at the height of one of the worst wildfire seasons in U.S. history shows vividly the extent of the disaster. The green triangles and red areas show active fires on August 19, 2000.

A satellite's-eye view

The GeoMAC site brings together real-time information—about the location and path of fires, as well as weather conditions—and combines it with basemaps that show fire managers in detail what kind of terrain and natural obstacles fire crews must contend with.

Green triangles indicate active fires reported from the field every morning and entered into a database maintained by the National Interagency Fire Center. From that, managers can get more detailed information via the Web about exactly how big the fire is, where it is, the number of people fighting it, the amount of damage it might end up causing, and possible causes.

Inset map can be turned on and off.

Oregon/Idaho border

Pink indicates urban areas; white dots are towns.

Red shapes delineate actively burning fires; orange, areas that burned within the previous 24 hours; and black, fires that burned 24 to 36 hours previously. This data comes from an infrared satellite that passes overhead and downlinks its information twice a day; sometimes it can show more about a fire's movements than people on the ground know, and can do so more quickly. The gray shapes indicate the perimeters of a fire as a whole, including previously burned areas.

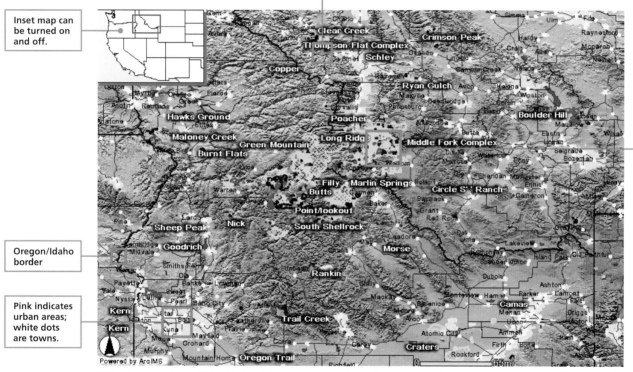

This August view of the Idaho/Montana area shows how active the fire season was, and how much detail the public GeoMAC Web site could show. All the layers available on the public site have been turned on in this view.

A new view of destruction

The GeoMAC site came online just as a bad fire season got worse: at its height, a million acres were estimated to be on fire, with more than thirty thousand people working in fire-related jobs. Such extreme conditions required extreme strategies and tactics; in some places, larger fires were simply allowed to burn out so resources could be moved to smaller fires that posed a bigger threat, or that offered a better chance of suppression.

Since managers could now see—literally—the entire picture, sometimes better than could those on the ground, decisions were made with more confidence. For example, the GeoMAC site allowed them to see that a group of small fires was consolidating into one large fire, burning away from an urban area. Therefore, firefighters could be moved elsewhere.

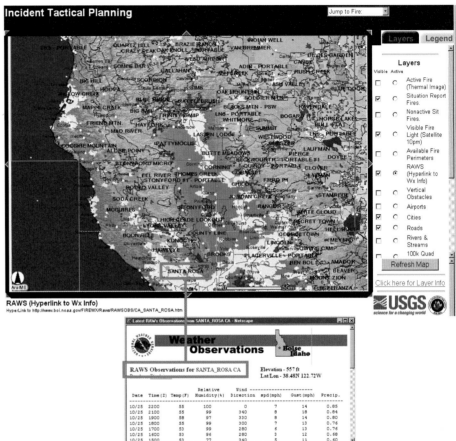

Members of the GeoMAC team can access additional resources and data layers at a nonpublic Web site. Above, clicking on a blue-dot hyperlink lets planners see real-time weather information from the National Weather Service, such as wind speed, direction, and humidity. The red squares indicate areas covered by available 1:250,000-scale USGS topographic maps. Wildfires and air traffic do not mix well, and the demonstration aviation site at the left shows how the locations of airports and restricted flight zones would be represented in a wildfire area.

Spreading the wealth and resources

GeoMAC is more than a Web site; it's an entirely new and powerful information resource for all firefighters, not just those at senior command levels.

Consolidating the GeoMAC data in one place has the additional benefit of giving commanders in the field—with flames roaring only a few feet away—an easier way to get the specific maps and data they need. Even though those commanders may not have the bandwidth to run the GeoMAC site at full throttle, its developers have set up data links that allow them, with a laptop and the right software, to download the specific data they need, even if it is from a database thousands of miles away, or from a satellite miles above the earth.

Think of it all as Pulaski, version 2.0.

The GeoMAC site is fueled by a broad array of government agencies providing a variety of data important in fire-fighting activities, all channeled through ArcSDE and ArcIMS software to the public and firefighters.

ArcIMS GeoMAC Web site

ArcSDE Oracle

Local Disk Storage

AML™ Data Processes

NOAA — Fuels Data / Lightning Data → Technical Specialists USGS/BLM/USFS ESRI ← Geospatial Archives — USGS EROS Data Center

AVHRR Thermal Locations

Aerial Hazards Data / Weather Data (RAWS)

USFS Region 4 Ogden

BLM NTSC Missoula Fire Lab

Base Data TOPO Maps/DRG's/ DQQQs/Watersheds

ORACLE Database Link ARC/INFO AML Processing

USGS/NDSI (National Spatial Data Infrastructure)

Field Observation Remotely Sensed GPS Perimeter Data

NIFC Situation Report Database Kansas City

Field Offices

The system

IBM Netfinity NT server with four 500-MHz hard drives and 2 GB RAM, and Sun Ultra™ Enterprise 450 running ArcIMS, ArcSDE, and Oracle8.

The URL

wildfire.usgs.gov

Acknowledgments

Thanks to Robert Plantrich, Bureau of Indian Affairs; John Guthrie and Dave Ozman, U.S. Geological Survey; Jeff Baranyi and Tim Clark, ESRI–Denver; and Sheri Ascherfeld, National Interagency Fire Center.

Online in Oregon

8

The list of communities across America that got walloped in the 1980s and 1990s by shifting economic and political tides is long and varied: Texas battered by gyrating crude oil prices; great metropolitan manufacturing engines rusting in the Midwest; chainsaws falling silent in the Pacific Northwest. Shifting tides like these mean finding new ways to swim.

In Douglas County, Oregon, timber drove the economy. Vast stretches of federal land within its borders provided logging and sawmill jobs for thousands, part of a regional economy responsible for billions of board-feet of lumber that built homes and businesses across America and the world.

But in the early 1990s, this economic engine began to backfire: administrative and court mandates imposed restrictions on logging on public lands, because of environmental concerns. There were labor problems, and a

growing awareness that the industry's resources were undeniably finite.

Local government agencies, including several towns and Douglas County itself, banded together under the Umpqua

Regional Council of Governments and in 1997 won a federal grant for a technology demonstration project that emphasized economic development.

A three-pronged attack

The project was designed to help communities identify and promote their own unique economic development opportunities. This goal of combining geography with information dissemination made GIS served over the Internet a natural choice.

Beyond these direct goals, the Douglas County project generated other benefits, including uniting different local governments toward a common goal.

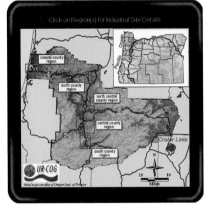

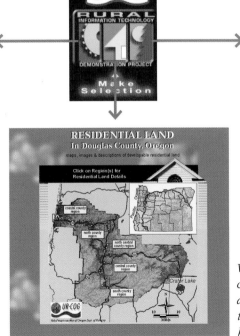

Visitors to the council's site are asked to choose from three Opportunity Areas: housing development, industrial development, and tourism. Each of these in turn has relevant hyperlinked sections in the county map.

Room to move

Douglas County is a place where you can still breathe air that you can't also see, and a place where there is still room to move. Home builders interested in selling this slower-paced lifestyle can assess the possibilities on the Housing Development Opportunities portion of the Web site.

Hyperlinked sections of the map pinpoint down to parcel level areas where both land and infrastructure are ready for residential development.

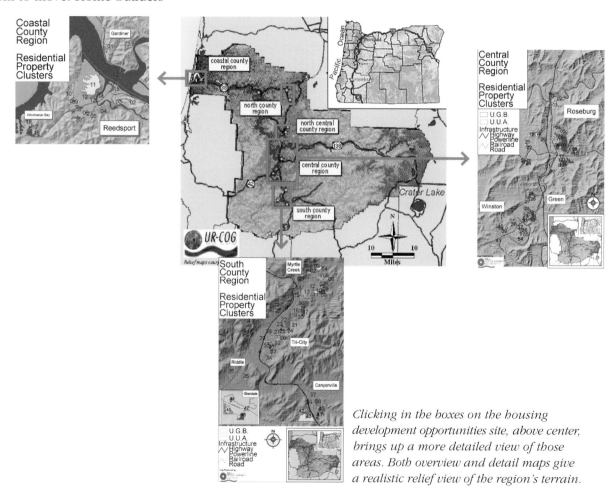

Clicking in the boxes on the housing development opportunities site, above center, brings up a more detailed view of those areas. Both overview and detail maps give a realistic relief view of the region's terrain.

Getting down to details

The site offers about as many ways to view a piece of available property as one could ever need. There is a basic map showing the detailed locations of residential clusters. Then, from the navigation bar at the top of that map, the visitor can inspect aerial photos of the same area, maps of zoning restrictions, or the locations of sewer, water, and power lines in relation to the housing tract.

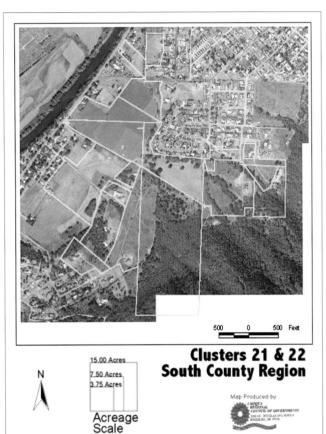

Clicking on the marked area of the previous page's maps brings up the property map at the right. The Site Map button at the top takes the viewer to an aerial photo of the same area.

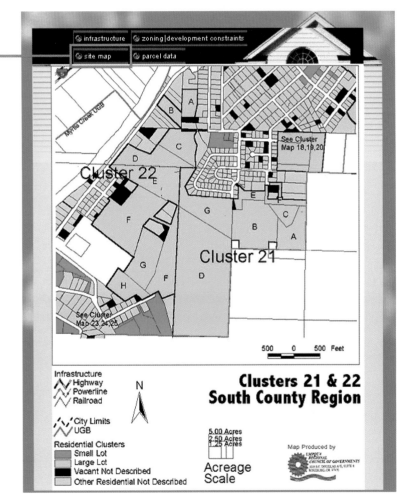

Industrial-strength information

Similarly, a mouse click on one of the site's Industrial Opportunity maps brings up a wealth of geographic information, which can influence decision making in a way a written real-estate report cannot.

Seeing in an aerial photo how close to the Umpqua River an industrial site is situated—as opposed to simply reading this bare fact in a report full of numbers—brings a new kind of reality to decision making, especially when that decision has to be made halfway across the country.

Any of these three views of the industrial opportunities in the Reedsport area are available by clicking within the boxed area (center). Written reports about the site, like the one at the lower right, are also available if the user really wants one.

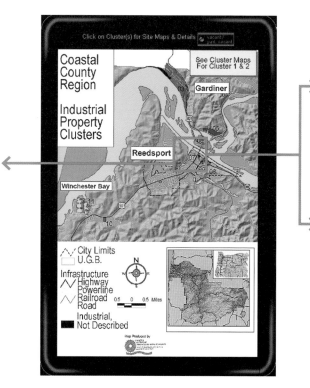

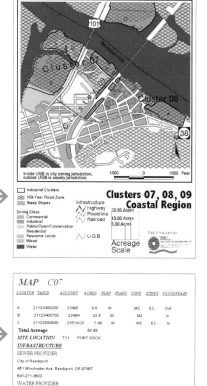

Getting grounded

Douglas County offers unparalleled opportunities for communing with nature. The tourism section of the Web site highlights the variety of those opportunities with relief maps that show terrain and vegetation, and that pinpoint the location of services that nature cannot provide.

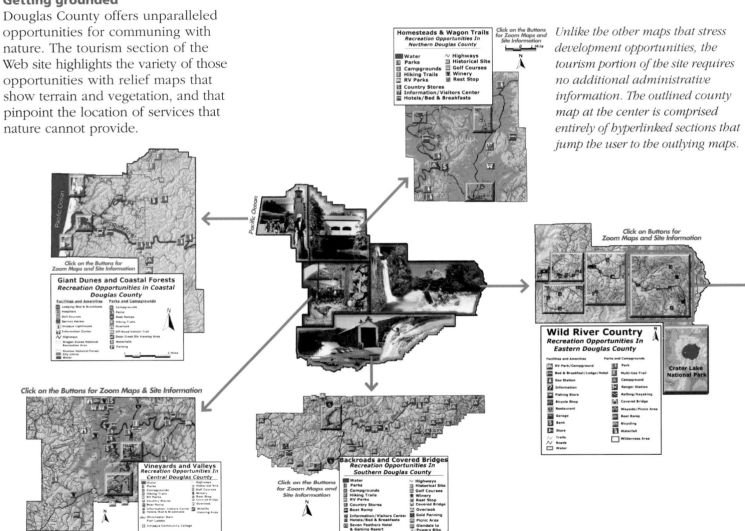

Unlike the other maps that stress development opportunities, the tourism portion of the site requires no additional administrative information. The outlined county map at the center is comprised entirely of hyperlinked sections that jump the user to the outlying maps.

Homesteads & Wagon Trails
Recreation Opportunities In Northern Douglas County

Click on the Buttons for Zoom Maps and Site Information

- Water
- Parks
- Campgrounds
- Hiking Trails
- RV Parks
- Country Stores
- Information/Visitors Center
- Hotels/Bed & Breakfasts
- Highways
- Historical Site
- Golf Courses
- Winery
- Rest Stop

Click on the Buttons for Zoom Maps and Site Information

Giant Dunes and Coastal Forests
Recreation Opportunities in Coastal Douglas County

Facilities and Amenities
- Lodging/Bed & Breakfasts
- Hospitals
- Golf Courses
- Salmon Harbor
- Umpqua Lighthouse
- Information Center
- Highways
- Oregon Dunes National Recreation Area
- Siuslaw National Forest
- City Limits
- Water

Parks and Campgrounds
- Campgrounds
- Parks
- Boat Ramps
- Hiking Trails
- Overlook
- Off-Road Vehicle Trail
- Dean Creek Elk Viewing Area
- Waterfalls
- Parking

Click on the Buttons for Zoom Maps & Site Information

Vineyards and Valleys
Recreation Opportunities In Central Douglas County

- Water
- Parks
- Campgrounds
- Hiking Trails
- RV Parks
- Country Stores
- Boat Ramp
- Information/Visitors Center
- Hotels/Bed & Breakfasts
- Seven Feathers Hotel & Gaming Resort
- Highways
- Historical Site
- Golf Courses
- Winery
- Covered Bridge
- Overlook
- Wildlife Viewing Area
- Winchester Dam Fish Ladder
- Umpqua Community College

Click on the Buttons for Zoom Maps and Site Information

Backroads and Covered Bridges
Recreation Opportunities In Southern Douglas County

- Water
- Parks
- Campgrounds
- Hiking Trails
- RV Parks
- Country Stores
- Boat Ramp
- Information/Visitors Center
- Hotels/Bed & Breakfasts
- Seven Feathers Hotel & Gaming Resort
- Rafting/Kayaking
- Swimming
- Highways
- Historical Site
- Golf Courses
- Winery
- Rest Stop
- Covered Bridge
- Overlook
- Gold Panning
- Picnic Area
- Glendale to Powers Bike Route
- Recreation Rentals

Click on Buttons for Zoom Maps and Site Information

Wild River Country
Recreation Opportunities In Eastern Douglas County

Facilities and Amenities
- RV Park/Campground
- Bed & Breakfast/Lodge/Hotel
- Gas Station
- Information
- Fishing Store
- Bicycle Shop
- Restaurant
- Garage
- Bank
- Store
- Trails
- Roads
- Water

Parks and Campgrounds
- Park
- Multi-Use Trail
- Campground
- Ranger Station
- Rafting/Kayaking
- Covered Bridge
- Wayside/Picnic Area
- Boat Ramp
- Bicycling
- Waterfall
- Wilderness Area

Crater Lake National Park

Rivers, trails, and places to rest

Clicking on the hyperlinked buttons within each area map brings up detailed location information about campground fees, the length and difficulty of hiking trails, the hours that wineries are open, and the myriad other necessities of a counterurban experience.

The Umpqua site is aided in its mission to promote tourism by other agencies, since the county itself has little direct responsibility for such activities, or for the kind of information tourists want. The photo of Toketee Falls, top, is homegrown, but most of the links are to outside agencies such as the U.S. Forest Service and the National Park Service, right, and to commercial sites. Any of these in turn can lead a visitor even deeper into uncharted territories of the cyberwilderness, beyond the safety of the Umpqua site.

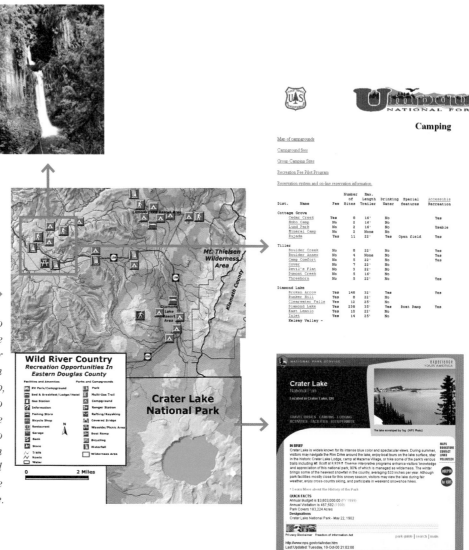

Camping

Map of campgrounds

Campground fees

Group Camping Sites

Recreation Fee Pilot Program

Reservation system and on-line reservation information

Dist. Name	Fee	Number of Sites	Max. Length Trailer	Drinking Water	Special features	Accessible Recreation
Cottage Grove						
Cedar Creek	Yes	8	16'	No		Yes
Hobo Camp	No	2	16'	No		
Lund Park	No	2	16'	No		Usable
Mineral Camp	No	2	None	No		
Rujada	Yes	11	22'	Yes	Open field	Yes
Tiller						
Boulder Creek	No	8	22'	No		Yes
Boulder Annex	No	4	None	No		Yes
Camp Comfort	No	5	22'	No		Yes
Cover	No	7	22'	No		
Devil's Flat	No	3	22'	No		
Dumont Creek	No	5	16'	No		
Threehorn	No	5	22'	No		Yes
Diamond Lake						
Broken Arrow	Yes	148	32'	Yes		Yes
Bunker Hill	Yes	8	22'	Yes		
Clearwater Falls	Yes	12	25'	No		
Diamond Lake	Yes	238	35'	Yes	Boat Ramp	Yes
East Lemolo	Yes	10	22'	No		
Inlet	Yes	14	25'	No		
Kelsay Valley	-					

Wild River Country
Recreation Opportunities In Eastern Douglas County

Facilities and Amenities
- RV Park/Campground
- Bed & Breakfast/Lodge/Hotel
- Gas Station
- Information
- Fishing Store
- Bicycle Shop
- Restaurant
- Garage
- Bank
- Store
- Trails
- Roads
- Water

Parks and Campgrounds
- Park
- Multi-Use Trail
- Campground
- Ranger Station
- Rafting/Kayaking
- Covered Bridge
- Wayside/Picnic Area
- Boat Ramp
- Bicycling
- Waterfall
- Wilderness Area

N

0   2 Miles

Mt. Thielsen Wilderness Area

Klamath County

Diamond Lake Resort Area

Crater Lake National Park

NATIONAL PARK SERVICE

experience YOUR AMERICA

Crater Lake
National Park
Located in Crater Lake, OR

TRAVEL BASICS CAMPING LODGING
ACTIVITIES FACILITIES ECOSERVICES

The lake enveloped by fog (NPS Photo)

IN BRIEF
Crater Lake is widely known for its intense blue color and spectacular views. During summer, visitors may navigate the Rim Drive around the lake, enjoy boat tours on the lake surface, stay in the historic Crater Lake Lodge, camp at Mazama Village, or hike some of the park's various trails including Mt. Scott at 8,929 ft. Diverse interpretive programs enhance visitors' knowledge and appreciation of this national park, 90% of which is managed as wilderness. The winter brings some of the heaviest snowfall in the country, averaging 533 inches per year. Although park facilities mostly close for this snowy season, visitors may view the lake during fair weather, enjoy cross-country skiing, and participate in weekend snowshoe hikes.

* Learn More about the History of the Park

QUICK FACTS
Annual Budget is $3,803,000.00 (FY 1999)
Annual Visitation is 467,592 (1999)
Park Covers 183,224 Acres
Designations
Crater Lake National Park - May 22, 1902

Privacy Disclaimer   Freedom of Information Act   park guide | search | main

http://www.nps.gov/crla/index.htm
Last Updated: Tuesday, 10-Oct-00 21:02:00

MAPS
DIRECTIONS
CONTACT
LINKS
STAY
VOLUNTEER

The system

450-MHz Dell PC with 19 GB RAM, running ArcView GIS 3.1 with ArcView Spatial Analyst and ArcView 3D Analyst™ extensions and ESRI Data Automation Kit.

Maps created in ArcView GIS, captured in Paint Shop Pro, and imported into the site as image files.

The URL

www.ur-cog.cog.or.us

Acknowledgments

Thanks to Chuck Perino and Jeff Willis of the Umpqua Regional Council of Governments, and to Eric Fladager, formerly of UR-COG, now with Sutherlin, Oregon.

9

A healthy place to call home

Once, the homegrown, hometown newspaper would be all that folks would need to keep tabs on even the smallest details of life in their little corner of the globe. But media consolidations and the rise of cable television have left many communities across America lacking such a resource, old-fashioned as it may have been; it was one way people could learn which local events and issues might affect their lives.

It is in this information vacuum that geographic e-government services may have their greatest impact. By delivering maps over the Internet that pinpoint the location of potential sources of trouble in a community—crime patterns, perhaps, or environmental hazards—governments can keep their constituents informed, at low cost, yet with great precision. At one federal agency, they have started doing just this.

The mission of that agency, the Department of Housing and Urban Development (HUD), is "to promote adequate and affordable housing, economic opportunity, and a suitable living environment without discrimination for all Americans." Over the years, HUD's efforts to help American communities remain vital and healthy have produced a wide variety of programs to reduce homelessness, to fight discrimination, to make home ownership easier, to make more police available in neighborhoods, and to revitalize abandoned industrial sites known as brownfields.

Online D.C.

Like many federal agencies, HUD has developed a strong e-government presence on the World Wide Web. From its home page, consumers can perform many tasks: find grant funds, file complaints, read the latest about HUD programs.

The neighborhood has seemed safer since Officer McCoy moved in, thanks to the Officer Next Door program

WEST MAIN

Move your pointer over objects in the scene above. Click on these objects for more information about HUD's programs.

HUD's Homes and Communities page, left, provides instant access through hyperlinks to the wide array of HUD resources. One of these, above, shows a fictional city block where each building has been improved in some fashion by a HUD program. Clicking on a building, such as the blue one where the cursor is located, brings up hyperlinked information about that program.

Consolidating resources

Perhaps the most powerful HUD hyperlink takes you to its Environmental Mapping service—its E-maps site.

HUD's E-maps can be drawn for any part of the country, with data from three agencies: HUD itself, the Environmental Protection Agency (EPA), and the Census Bureau.

HUD's maps help you find HUD-subsidized low-income housing, the location of brownfields revitalization zones, and the locations of other HUD programs.

The EPA data brings together all of that agency's information about the location and nature of environmental data, such as hazardous waste sites.

Census data lets you create detailed maps of Americans themselves—who they are, where they live, and how they live.

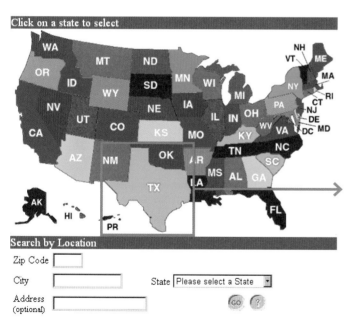

Search for a location by ZIP code, city and state, or address.
Zoom into an area to see HUD and EPA projects.

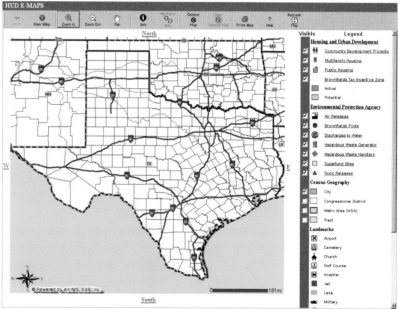

Creating an environmental and housing map of your community begins with the map at the left, choosing state, city, ZIP Code, or any combination. In this case, clicking on Texas brings up an E-map of that state, right.

Home is where the map is

Finding HUD programs, for example, is as easy as drawing a rectangle around any area of the United States. HUD's computers zoom in on that area while searching the huge E-maps database for programs and exact locations for the area.

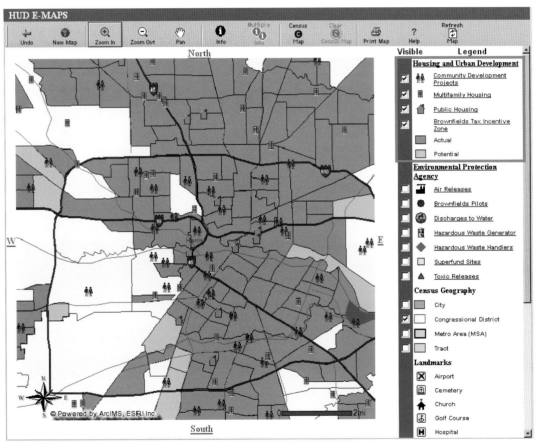

The tools above the map display let the user manipulate the map. The list at the right contains the legend for the features on the map, and also the controls for choosing which layer to turn on or off. On this map, only the HUD features are turned on, showing the location of public housing and HUD programs in the Houston area.

Just the facts

The environment became a hot political issue during the 2000 campaign for U.S. president, with the candidates arguing about the effectiveness of some environmental programs. The HUD site was the place voters could go for the facts.

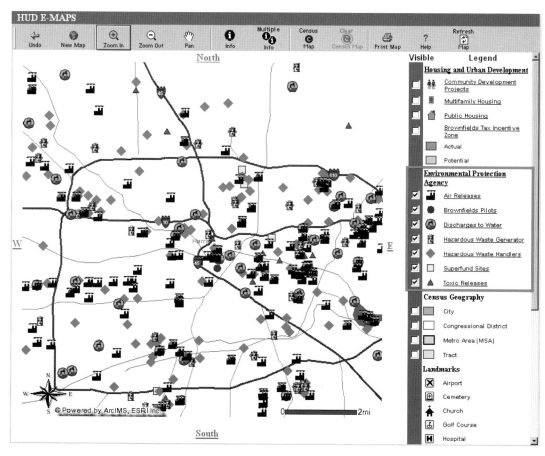

HUD E-MAPS

Visible	Legend
	Housing and Urban Development
☐	Community Development Projects
☐	Multifamily Housing
☐	Public Housing
☐	Brownfields Tax Incentive Zone
	Actual
	Potential
	Environmental Protection Agency
☑	Air Releases
☑	Brownfields Pilots
☑	Discharges to Water
☑	Hazardous Waste Generator
☑	Hazardous Waste Handlers
☑	Superfund Sites
☑	Toxic Releases
	Census Geography
☐	City
☐	Congressional District
☐	Metro Area (MSA)
☐	Tract
	Landmarks
	Airport
	Cemetery
	Church
	Golf Course
	Hospital

The same area of Houston, with HUD features turned off and EPA features turned on, shows the distribution of hazardous waste generators, Superfund toxic-waste sites, and other potentially hazardous locations throughout the area.

A portrait in numbers

With census data from the E-maps page, you can combine demographics with the HUD and EPA data.

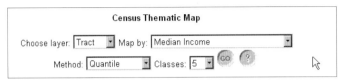

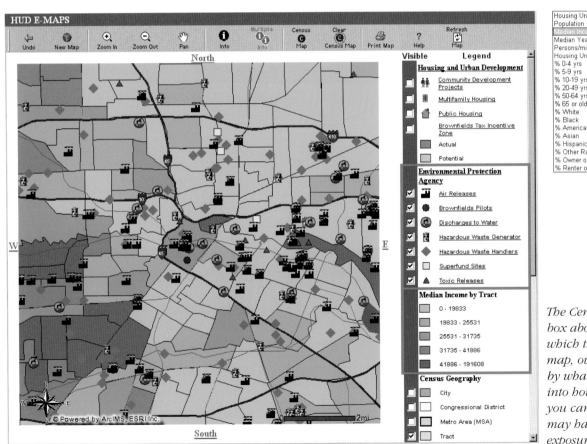

The Census tool brings up the dialog box above, which lets a visitor select which theme (median income) to map, over what geography (tract), by what method (quantile), and into how many levels (five). Here, you can judge which income groups may have more hazardous-waste exposure.

Digging deeper

Environmental data from the EPA is widely available on the Web. It is also available directly from the EPA's Enviromapper service Web page. But whether from the HUD site or the EPA site, the visitor has access to several levels of data beyond the initial look at location, or location in a demographic context: clicking the appropriate links from a symbol on a GIS map will take you to the full text of public EPA reports. With enough digging, you can find almost every conceivable piece of information about a source of environmental concern.

As well as depth, it is the range of information resources that the HUD E-maps site offers that helps it satisfy the numerous promises of e-government: housing assistance can be located; chemicals and hazardous materials can be pinpointed down to the street level; the demographics of an area, through census figures, can be visualized; and all can be combined on a map to clarify relationships among all those sources of data.

It is sometimes said that the mark of intelligence is not the quantity of facts you know, but your ability to see relationships among those facts. In helping visualize such relationships, the E-maps site takes a giant step toward creating some very intelligent consumers of government services.

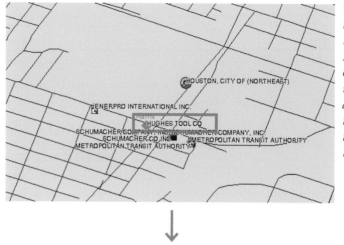

The Identify tool on the E-maps page leads from the Houston location of one hazardous waste handler to an EPA report on the exact nature of the material being handled.

Toxic Releases for Reporting Year 1992

TRI FACILITY ID: 77023HGHST5425P

SIC Codes for 1992

SIC CODE	SIC CODE DESCRIPTION
3533	OIL AND GAS FIELD MACHINERY AND EQUIPMENT

Chemicals Transferred to other Sites

CHEMICAL NAME	TRI CHEM ID	DOCUMENT	RELEASE AMOUNTS LBS/YR	RELEASE BASIS CODE	TYPE OF WASTE MANAGEMENT	TRANSFER SITE NAME	TRANSFER SITE CITY
MANGANESE	007439965	1392060261601	85	OTHER	METALS RECOVERY	INDUSTRIAL RECOVERY SERVICE	GARLAND
MANGANESE	007439965	1392060261601	170	OTHER	LANDFILL/DISPOSAL SURFACE IMPOUNDMENT	BROWNING FERRIS INDUSTRIES HOU, STON LANDFILL DISTICT	HOUSTON
MANGANESE	007439965	1392060261601	21000	OTHER	METALS RECOVERY	PROLER INTERNATION CORP.	HOUSTON
METHANOL	000067561	1392060261599	370	MASS BALANCE CALCULATIONS	REUSE AS FUEL/FUEL BLENDING	SAFETY-KLEEN CORP.	MISSOURI CUTY
NICKEL	007440020	1392060261587	1	MASS BALANCE CALCULATIONS	TRANSFER TO WASTE BROKER - ENERGY RECOVERY	HOUSTON FUEL OIL TERMINAL CO.	HOUSTON

The system

The E-maps application uses 700-MHz PowerEdge 6400 Xeon servers, each with 512 MB RAM, and six 9.1-GB RAID-5 hard drives. Software includes ArcIMS, ArcSDE, Oracle8, Microsoft Windows® 2000, and Microsoft Internet Information Server.

The URL

www.hud.gov/emaps

Acknowledgments

Thanks to Todd Rogers, ESRI–Washington, D.C., and to Dave Nystrom, Department of Housing and Urban Development.

10

E-city, e-community

Thousands once rushed to California to mine rich lodes of gold, but a century and a half later, information has become the world's most sought-after commodity. In Sacramento, capital of the Golden State, GIS and the Internet are the tools of choice for mining this twenty-first-century ore. Easy-to-use interactive maps, named Map Information Tools, on the Sacramento city Web site, let residents wander electronically through the city and its neighborhoods at will. With only a mouse click, they can track down details of government services, and of the city's character: zoning, schools, crime patterns, property values, neighborhood centers, and community services.

Several factors may explain Sacramento's advanced way of disseminating information. As home to the core government functions of the nation's most populous state, Sacramento hardly suffers from an information shortage;

processing information, after all, is what government is all about. Sacramento is also a long-time user of GIS technology, having had a variety of mapping products on its computers, in the police and planning departments, since 1991. Add to this

a tradition of citizen involvement— a city program to train interested residents in the intricacies of municipal government was recognized by the American Society for Public Administration with an Outstanding Innovation in Government award.

Framing the city

Sacramento's experience shows how easily any agency can put GIS and the Internet to work. Built with MapObjects Internet Map Server (IMS) by a one-woman GIS development department, the site uses standard ESRI-built templates for local government maps. These templates provide a single, robust framework on which to deliver many kinds of information.

The site divides geographic information about the city into four subject areas:

- General maps, which show city council districts, neighborhood service areas, zoning, and other desiderata.
- Property maps, which show individual parcel data supplied by the county tax assessor.
- Crime maps.
- Code enforcement maps, which show the locations of conditions inimical to a healthy and safe community, such as abandoned buildings and cars, zoning violations, and graffiti.

Not only does the common template make Web site design easier, it also gives the entire site a common look and feel, easing the process of navigation for city residents who may be new to interactive mapping.

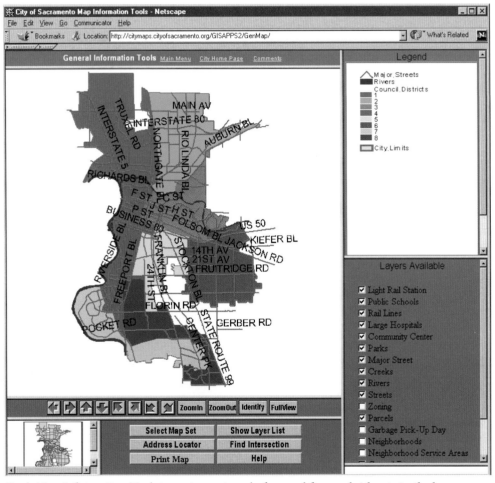

Each Map Information Tools page is composed of several frames laid out similarly: a large, central frame where the main digital map of Sacramento resides; smaller frames containing tools for the user to work in the main map, including changing its size and level of visible detail; and a legend frame to keep the user understanding exactly what those things are in the main map.

Multiple perspectives

From the general maps page, users can descend the city's informational mineshafts in two ways. Via the Select Map Set button, the user brings up a map of one of several complete themes of the city: council districts, neighborhood service areas, neighborhoods, garbage pick-up days, and zoning.

Another choice is for a map the user defines. Choosing this option lets you build custom maps using one or more of seventeen different thematic layers, among them hospitals, streets, schools, parks, and the routes of Sacramento's streetcar lines.

These can be combined any way you want, allowing you to find relationships among features—the number of parks in a particular council district, for example, or a school's proximity to busy streets.

As has become common in digital mapping applications, some map layers do not become visible until you have zoomed in to a smaller scale. This prevents the screen from becoming too cluttered when a general view of the city is chosen.

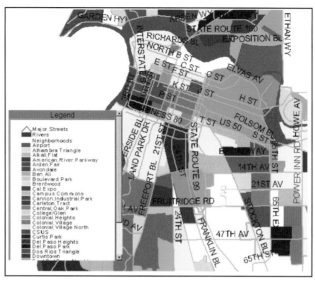

A map of Sacramento neighborhoods, accessible from the Select Map Set button.

Also from the Select Map Set button, residents can see combined map layers. This is a portion of the zoning at the border of the 4th and 5th City Council districts.

Just the maps, ma'am

The most visited pages on the Sacramento site are those that feature the locations of crime in the city.

As with the general map pages, you can choose from a citywide map of a particular kind of crime, or mix and match crimes to show differences and similarities of the city's crime patterns.

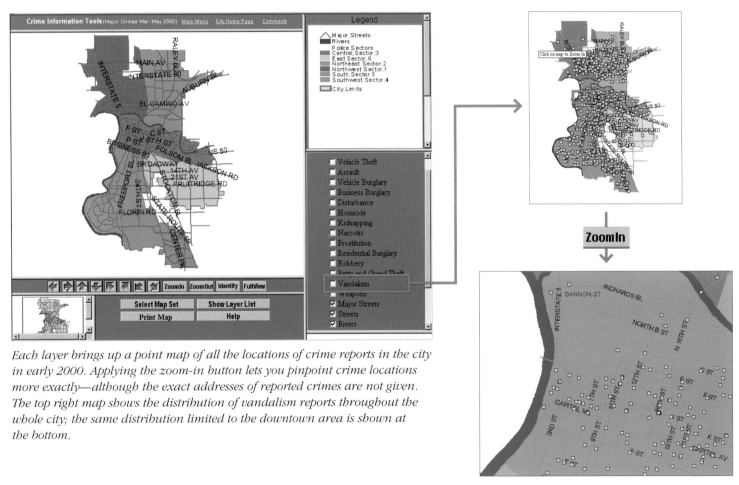

Each layer brings up a point map of all the locations of crime reports in the city in early 2000. Applying the zoom-in button lets you pinpoint crime locations more exactly—although the exact addresses of reported crimes are not given. The top right map shows the distribution of vandalism reports throughout the whole city; the same distribution limited to the downtown area is shown at the bottom.

A tool for citizen action

Generally considered less serious than felonies or misdemeanors, violations of ordinances, administrative rules, zoning laws, and health codes are nevertheless something a city wants to minimize within its boundaries. These kinds of violations—undrivable cars gathering cobwebs in the street, for example—may lead to a broader deterioration in the quality of life in a neighborhood, and consequently in the city.

Sacramento's code violation maps can show residents and officials quickly and easily the locations and patterns of such problems, so they can more easily take action.

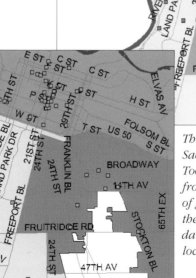

Three views of the same area of Sacramento from the Code Violations Tools page show the pattern of problems from April to June 2000. Locations of graffiti are shown at the left; in the center, locations of abandoned or dangerous buildings; and at the top, locations of abandoned vehicles.

The value of property

For most people a home is the biggest single investment of their lives. Questions about their property's value, and the taxes they must pay, therefore, are ones they take very seriously.

Recognizing this, Sacramento, like many other jurisdictions, has made sure that homeowners have easy access to information about their property. The Map Information Tools pages give residents three ways of mapping the property they own.

They can zoom in on the property from the main map; they can enter their parcel number, if they know it; or they can enter their street address. Any of the three methods pinpoints the property and brings up detailed information, such as the exact amount of property taxes due.

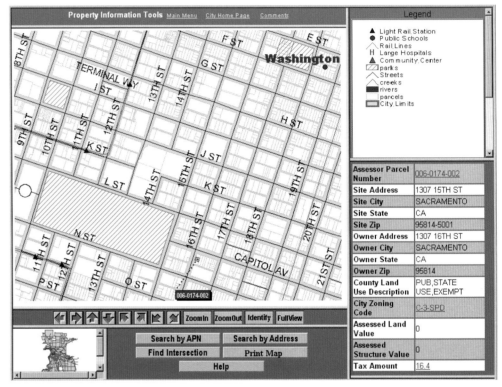

In this example, a state-owned building in downtown Sacramento near the Capitol is highlighted. The information screen at the right indicates the property is exempt from taxes because it is a public building.

Taking care of the basics

While Sacramento's map pages are filled with enough information to fill a dissertation on urban planning, many residents want to know more immediate, if mundane, information: where to get a dog license, where to go to throw a softball, how far do the kids have to walk to school? For these simple tasks, the Sacramento site is admirably suited. Schools and parks, for instance, can easily be seen from the General Maps page. Additional, similarly practical tools include Locator and Find Intersection buttons, which, when pressed, bring you to the exact location within the city that you're looking for.

If only gold had been as easy to find.

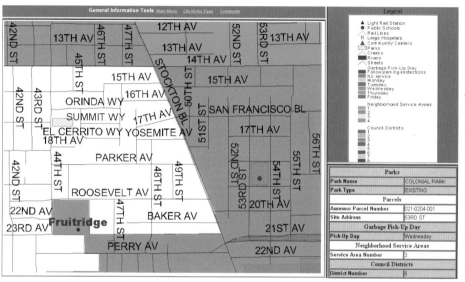

Using the Identify button to find information about a park in a southeastern section of Sacramento can give you much more. At left, a visitor learns not only the name of the park, but the council district it is in, the days when trash is picked up in the area, and the location and names of nearby schools. Below, finding the intersection of O St. and 15th St. also shows the route of the city's streetcar line.

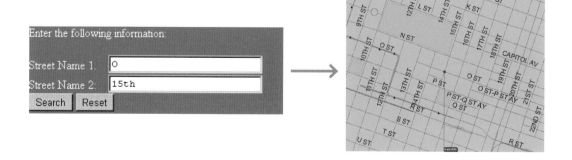

The system

Compaq Proliant 1850R 6/500 with 256 MB RAM and 9.1-GB hard drive, running MapObjects IMS.

The URL

citymaps.cityofsacramento.org

Acknowledgments

Thanks to Maria MacGunigal, GIS project manager, City of Sacramento.

Worth a look

The definition of e-government continues to expand, with city managers, city council members, state representatives, GIS specialists, Web developers, and others finding new ways each day to put interactive maps into the hands of the people. This chapter provides some snapshots of the different directions e-government interactive mapping is heading.

EnviroMapper

www.epa.gov/enviro/html/em

The Environmental Protection Agency's interactive mapping site is connected to a voluminous database of information about environmental hazards and potential hazards in every neighborhood in the country. Easy to navigate and understand, the site's Open Link feature helps local government Web sites connect to the site and display its maps.

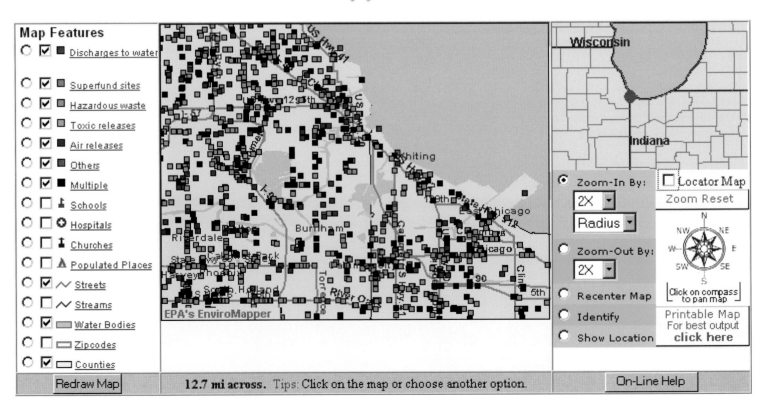

California Rivers Assessment

www.endeavor.des.ucdavis.edu/newcara

The California Rivers Assessment site provides a unique interface for navigating every river and river basin in California, giving anyone with an interest or stake in the health of those rivers access to ninety-nine data sets about water quality, dams, irrigation, endangered species, and much more. Originally funded by the California Resources Agency through the Wildlife Conservation Board, the site's builder and owner is the Information Center for the Environment (ICE) at the University of California, Davis. The center maintains an environmental information portal of notable breadth at ice.ucdavis.edu.

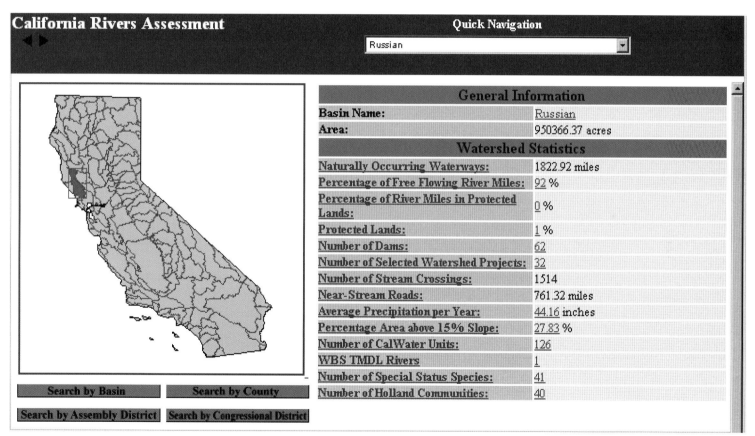

California Rivers Assessment

Quick Navigation

Russian

General Information	
Basin Name:	Russian
Area:	950366.37 acres
Watershed Statistics	
Naturally Occurring Waterways:	1822.92 miles
Percentage of Free Flowing River Miles:	92 %
Percentage of River Miles in Protected Lands:	0 %
Protected Lands:	1 %
Number of Dams:	62
Number of Selected Watershed Projects:	32
Number of Stream Crossings:	1514
Near-Stream Roads:	761.32 miles
Average Precipitation per Year:	44.16 inches
Percentage Area above 15% Slope:	27.83 %
Number of CalWater Units:	126
WBS TMDL Rivers	1
Number of Special Status Species:	41
Number of Holland Communities:	40

Search by Basin   Search by County

Search by Assembly District   Search by Congressional District

City of Hampton, Virginia

www.hampton.va.us

Hampton's interactive mapping site lets visitors combine disparate kinds of information about property into custom, compact views. Here, details about the parcel at the right—school districts, census tract number, political divisions—have been coupled with a map showing average noise levels in decibels from planes using nearby Langley Air Force Base.

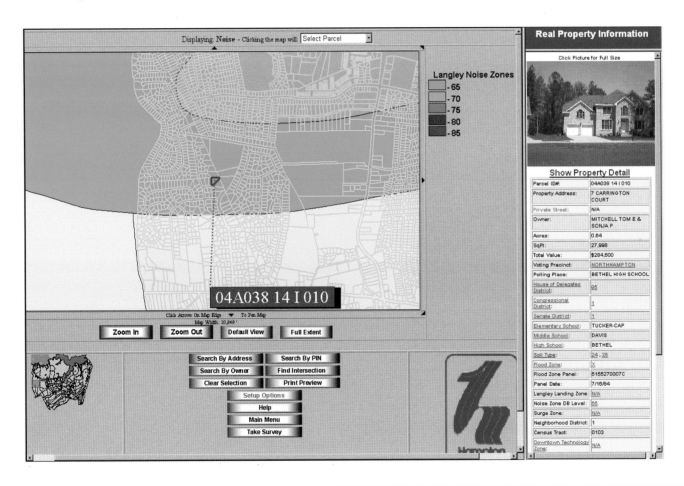

Hillsborough County, Florida
www.hcpafl.org

The Hillsborough County Property Appraiser site, encompassing Tampa and environs, features not only a sleek interface, but also a video air show. Clicking on the small airplane symbols, on Davis Island (below) and in downtown Tampa, initiates your desktop media player, showing an aerial flyby of the area in question.

Los Angeles, California

gis.lacity.org/routemap/RAP/data/RecandPark.htm

The City of Los Angeles, which uses MapObjects IMS for property applications, also uses *Route*MAP IMS to help residents find nearby places to play tennis, golf, swim, or do a number of other things. The application also creates a map of the best route to take. Below, recreation facilities within five miles of the city's Silverlake district are shown.

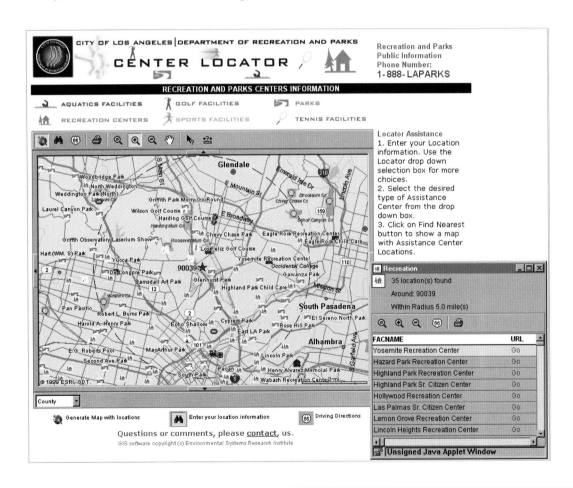

La Dolce Vita Cartografica

193.70.162.77/image/Zone.htm

The Italian penchant for interesting, even provocative design extends to the World Wide Web. The city of Rome's interactive mapping page can take you from a continental view of Europe to a street-level view of the city. Below, the browser is zoomed in on an area near Vatican City. The light green dots indicate the locations of embassies.

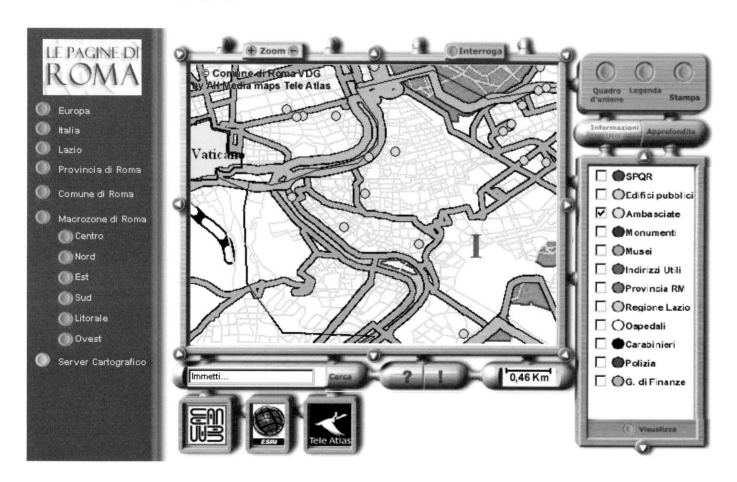

British Columbia

www.fishwizard.com

An abundance of natural resources makes British Columbia one of North America's most interesting environments. The province's Department of Fisheries uses interactive maps to show how one part of that abundance is distributed through thousands of miles of lakes, streams, and rivers. A section of the Fraser River watershed is shown below.

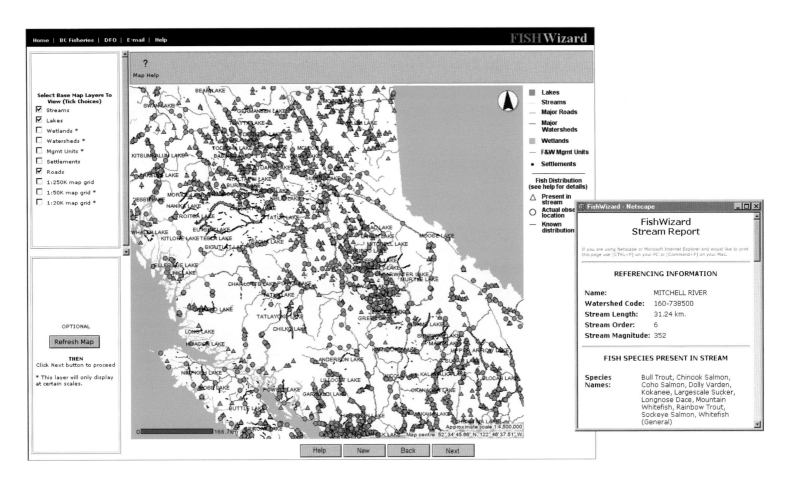

United States

bam.geoinsight.com

Despite little publicity, collisions with birds have caused millions of dollars of damage to civilian and military aircraft. The Air Force and Geo Insight International, Inc., developed a spatial model to help aviators find hazardous airspace, then mapped it online. Below, dangerous areas near three California air force bases in a two-week period in January, 2001.

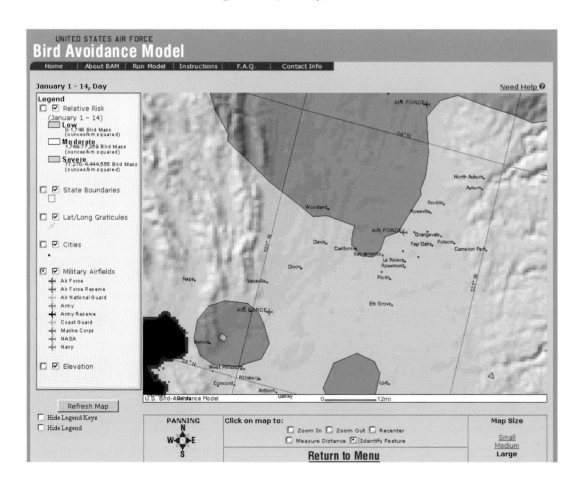

Geography Network and e-government

If you drew points on a map of all the e-government sites using IMS technology to serve their constituents—those we've detailed in this book and the many we didn't have room for—you'd have a patchwork: a point in Umpqua County, Oregon, another in Yavapai County, Arizona, a big blob in the San Diego region, more points in Tacoma and Delaware and Ohio, and others in hundreds of other places across the continent. Looking at such a map, you might then figure that some day, when all those dots got connected, you'd really have something: an awesome network covering the entire continent map with the power of interactive GIS mapping.

In fact, that day has already arrived.

The fabric on which all these dots are being stitched into one solid geographic quilt is Geography Network. An Internet portal located at www.geographynetwork.com℠, it's a kind of virtual geographic bazaar where you can find free and commercial geographic data of any kind for any part of the world: a place where the owners of maps and data, the providers of mapping and other geoservices, consumers and traders, can all meet and do business.

Agencies moving to e-government services will find their task both eased and strengthened by participation in Geography Network; it is there they will find other agencies with the data and services they need, and which in turn will need their data and their services.

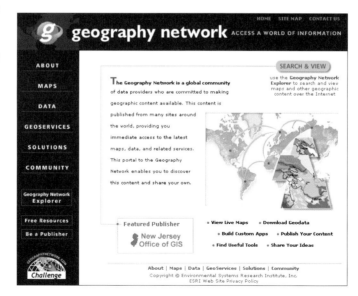

Connecting to the future

Geography Network is much more than a device for exchanging hyperlinks with other agencies and then hoping people will be able to find yours. It is rather a virtual community where agencies can make connections, among themselves and with consumers, that they might never have been aware were possible.

Through Geography Network, for instance, you might be able to find that obscure but crucial piece of basemap data that puts a new perspective on your part of the world—a perspective that a major manufacturer needs to decide where to site a new plant that would employ thousands of area residents.

Or you might use Geography Network to direct more people, including constituents, to your own site, giving you more bang for your Web buck.

The range of resources is expansive: there are static maps and dynamic, interactive maps, from both commercial and government providers—also known as publishers—and much of it for free. Many publishers, such as ESRI, also make the data for maps available separately for you to download and incorporate locally.

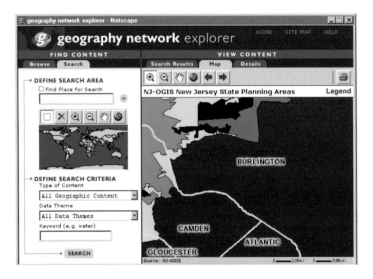

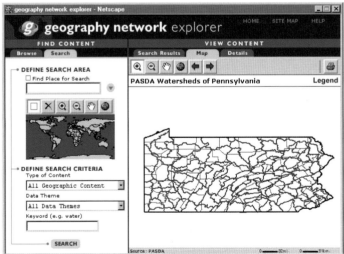

Maps and data sets for different features of New Jersey and Pennsylvania are available on Geography Network.

Services and solutions

Solutions to specific geographic problems are also available through Geography Network. You could use the portal to provide your own public services, perhaps routing students to their neighborhood schools, or buffering around particular locations.

Being a part of these new synergies is easy: start at the "Be a Publisher" page and fill out some simple online forms.

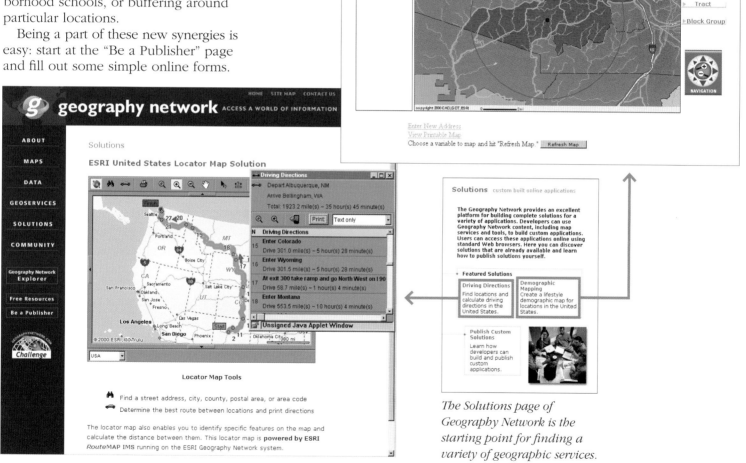

The Solutions page of Geography Network is the starting point for finding a variety of geographic services.

From ArcView to ArcIMS

The software that makes Geography Network and its ideas possible is ArcIMS, the latest IMS technology from ESRI.

Earlier map-serving technology continues to serve many agencies well, although ArcIMS will become the standard. With ArcView IMS, for instance, city agencies such as TacomaSpace got an easy-to-use extension to the basic, desktop ArcView GIS software that has been the mainstay of many agencies for years. It is called an extension because it does just that: it extends maps created in the ArcView environment to users out on the Internet.

The next evolution of IMS technology, MapObjects IMS, also continues to serve its customers well, including several highlighted in this book. More customizable than ArcView IMS because it was built on an entirely different architecture, MapObjects IMS requires more complex technical knowledge of the Windows® environment to install and operate, but has commensurately more power.

ArcIMS incorporates the best features of the earlier technologies, but is in fact the result of a new IMS architecture. Although completely different, its basic structure is not difficult to understand. Moreover, its clear, wizard-driven interfaces make it possible to set up a complete ArcIMS service on one machine within a few hours.

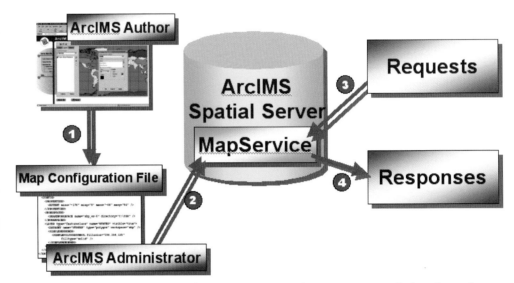

A geospatial e-government service begins in ArcIMS with a component called Author, where you choose the size, look, and scope of the map you want your Internet site's visitors to see. Author then writes your instructions to a special map configuration file. Another software component, Administrator, takes the information from that file and uses it to define how those maps will be made available to your site's visitors. Administrator is also where you set parameters for service demand and delivery; if you expect thousands of visitors per hour demanding maps and data queries, it is this component that makes ArcIMS powerful enough to handle such demand. The third component, not shown here, called Designer, helps you create the actual Web pages containing the maps that your visitors will see. Designer's process uses templates that help you decide the kinds of tools to make available to visitors to your site—tools for tasks such as measuring distances, querying data, or identifying features such as property.

Networked to the world

Geography Network is based on ArcIMS technology, but that doesn't mean you need a full ArcIMS installation to take advantage of it. You can view maps and use services directly from www.geographynetwork.com, of course. Or you can use a free software tool from ESRI called ArcExplorer™ 3 Java™ Edition to connect to Geography Network.

ArcExplorer 3 Java Edition lets you connect with Geography Network map sites that provide feature services, so-called because they stream map features in real time over the Internet to you. This means that when you use ArcExplorer 3 to pan to a new location or to identify a feature, those tasks are done by the server on the distant Web site. Because that distant server is also usually where all the data is located, the ArcExplorer 3 viewer gives you data and processing power far beyond what you probably have available on your desk.

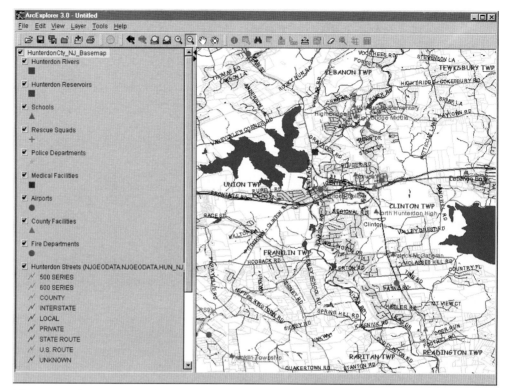

This ArcExplorer 3 view of a Geography Network e-government participant, New Jersey's Hunterdon County, shows the amount of detailed information available with this free software tool. The check marks in the boxes indicate which features are available for viewing, and the buttons at the top of the interface allow you to perform a variety of analytical tasks.

Integrating Geography Network

Geography Network's power is expected to be so widespread that accessing it has been made a part of the core functionality of ArcGIS, the newest GIS technology from ESRI.

One reason for this integration is the convergence that Geography Network represents—the convergence of GIS power and Internet power in a way that will reshape how we obtain and use geographic resources.

With its power distributed across the Internet through Geography Network's portal—with data for any location available from any location—GIS itself will shed many traditional limitations, including those having to do with cost and with data storage.

And with that easier, better access, governments offering e-government services will also shed traditional limitations on how they do their primary job—serving the public.

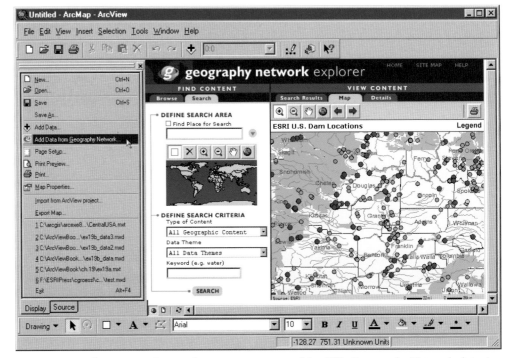

In ArcMap 8, the map-making software component of ArcGIS, Geography Network data is accessible from a drop-down menu just as any traditional data source is.

GIScience

GIS for Everyone SECOND EDITION
Now everyone can create smart maps for school, work, home, or community action using a personal computer. This revised second edition includes the ArcExplorer™ geographic data viewer and more than 500 megabytes of geographic data. ISBN 1-879102-91-9

The ESRI Guide to GIS Analysis
An important new book about how to do real analysis with a geographic information system. *The ESRI Guide to GIS Analysis, Volume 1: Geographic Patterns and Relationships* focuses on six of the most common geographic analysis tasks. ISBN 1-879102-06-4

Modeling Our World
With this comprehensive guide and reference to GIS data modeling and to the new geodatabase model introduced with ArcInfo™ 8, you'll learn how to make the right decisions about modeling data, from database design and data capture to spatial analysis and visual presentation. ISBN 1-879102-62-5

Hydrologic and Hydraulic Modeling Support with Geographic Information Systems
This book presents the invited papers in water resources at the 1999 ESRI International User Conference. Covering practical issues related to hydrologic and hydraulic water quantity modeling support using GIS, the concepts and techniques apply to any hydrologic and hydraulic model requiring spatial data or spatial visualization. ISBN 1-879102-80-3

Beyond Maps: GIS and Decision Making in Local Government
Beyond Maps shows how local governments are making geographic information systems true management tools. Packed with real-life examples, it explores innovative ways to use GIS to improve local government operations. ISBN 1-879102-79-X

The ESRI Press Dictionary of GIS Terminology
The *ESRI Press Dictionary of GIS Terminology* brings together the language and nomenclature of the many GIS-related disciplines and applications. Designed for students, professionals, researchers, and technicians, the dictionary provides succinct and accurate definitions of more than a thousand terms. ISBN 1-879102-78-1

ESRI Map Book:
Applications of Geographic Information Systems
A full-color collection of some of the finest maps produced using GIS software. Published annually since 1984, this unique book celebrates the mapping achievements of GIS professionals. ISBN 1-879102-83-8

CONTINUED ON NEXT PAGE

Other books from **ESRI Press** continued

The Case Studies Series

ArcView GIS Means Business
Written for business professionals, this book is a behind-the-scenes look at how some of America's most successful companies have used desktop GIS technology. The book is loaded with full-color illustrations and comes with a trial copy of ArcView GIS software and a GIS tutorial. ISBN 1-879102-51-X

Zeroing In:
Geographic Information Systems at Work in the Community
In twelve "tales from the digital map age," this book shows how people use GIS in their daily jobs. An accessible and engaging introduction to GIS for anyone who deals with geographic information. ISBN 1-879102-50-1

Serving Maps on the Internet
Take an insider's look at how today's forward-thinking organizations distribute map-based information via the Internet. Case studies cover a range of applications for ArcView Internet Map Server technology from ESRI. This book should interest anyone who wants to publish geospatial data on the World Wide Web. ISBN 1-879102-52-8

Managing Natural Resources with GIS
Find out how GIS technology helps people design solutions to such pressing challenges as wildfires, urban blight, air and water degradation, species endangerment, disaster mitigation, coastline erosion, and public education. The experiences of public and private organizations provide real-world examples. ISBN 1-879102-53-6

Enterprise GIS for Energy Companies
A volume of case studies showing how electric and gas utilities use geographic information systems to manage their facilities more cost effectively, find new market opportunities, and better serve their customers. ISBN 1-879102-48-X

Transportation GIS
From monitoring rail systems and airplane noise levels, to making bus routes more efficient and improving roads, this book describes how geographic information systems have emerged as the tool of choice for transportation planners. ISBN 1-879102-47-1

GIS for Landscape Architects
From Karen Hanna, noted landscape architect and GIS pioneer, comes *GIS for Landscape Architects*. Through actual examples, you'll learn how landscape architects, land planners, and designers now rely on GIS to create visual frameworks within which spatial data and information are gathered, interpreted, manipulated, and shared. ISBN 1-879102-64-1

GIS for Health Organizations
Health management is a rapidly developing field, where even slight shifts in policy affect the health care we receive. In this book, you'll see how physicians, public health officials, insurance providers, hospitals, epidemiologists, researchers, and HMO executives use GIS to focus resources to meet the needs of those in their care. ISBN 1-879102-65-X

GIS in Public Policy

This book shows how policy makers and others on the front lines of public service are putting GIS to work—to carry out the will of voters and legislators, and to inform and influence their decisions. *GIS in Public Policy* shows vividly the very real benefits of this new digital tool for anyone with an interest in, or influence over, the ways our institutions shape our lives. ISBN 1-879102-66-8

Integrating GIS and the Global Positioning System

The Global Positioning System is an explosively growing technology. *Integrating GIS and the Global Positioning System* covers the basics of GPS and presents several case studies that illustrate some of the ways the power of GPS is being harnessed to GIS, ensuring, among other benefits, increased accuracy in measurement and completeness of coverage. ISBN 1-879102-81-1

GIS in Schools

GIS is transforming classrooms—and learning—in elementary, middle, and high schools across North America. *GIS in Schools* documents what happens when students are exposed to GIS. The book gives teachers practical ideas about how to implement GIS in the classroom, and some theory behind the success stories. ISBN 1-879102-85-4

Disaster Response: GIS for Public Safety

GIS is making emergency management faster and more accurate in responding to natural disasters, providing a comprehensive and effective system of preparedness, mitigation, response, and recovery. Case studies include GIS use in siting fire stations, routing emergency response vehicles, controlling wildfires, assisting earthquake victims, improving public disaster preparedness, and much more. ISBN 1-879102-88-9

CONTINUED ON NEXT PAGE

Other books from **ESRI Press** continued